NEEM

A Tree For Solving Global Problems

Report of an Ad Hoc Panel of the
Board on Science and Technology for International Development
National Research Council

NATIONAL ACADEMY PRESS
Washington, D.C. 1992

NOTICE: The project that is the subject of this report was approved by the Governing Board of the National Research Council, whose members are drawn from the councils of the National Academy of Sciences, the National Academy of Engineering, and the Institute of Medicine. The members of the committee responsible for the report were chosen for their special competence and with regard for appropriate balance.

This report has been reviewed by a group other than the authors according to procedures approved by a Report Review Committee consisting of members of the National Academy of Sciences, the National Academy of Engineering, and the Institute of Medicine.

The National Academy of Sciences is a private, nonprofit, self-perpetuating society of distinguished scholars engaged in scientific and engineering research, dedicated to the furtherance of science and technology and to their use for the general welfare. Upon the authority of the charter granted to it by the Congress in 1863, the Academy has a mandate that requires it to advise the federal government on scientific and technical matters. Dr. Frank Press is president of the National Academy of Sciences.

The National Academy of Engineering was established in 1964, under the charter of the National Academy of Sciences, as a parallel organization of outstanding engineers. It is autonomous in its administration and in the selection of its members, sharing with the National Academy of Sciences the responsibility for advising the federal government. The National Academy of Engineering also sponsors engineering programs aimed at meeting national needs, encourages education and research, and recognizes the superior achievements of engineers. Dr. Robert M. White is president of the National Academy of Engineering.

The Institute of Medicine was established in 1970 by the National Academy of Sciences to secure the services of eminent members of appropriate professions in the examination of policy matters pertaining to the health of the public. The Institute acts under the responsibility given to the National Academy of Sciences by its congressional charter to be an adviser to the federal government and, upon its own initiative, to identify issues of medical care, research, and education. Dr. Stuart Bonderant is acting president of the Institute of Medicine.

The National Research Council was organized by the National Academy of Sciences in 1916 to associate the broad community of science and technology with the Academy's purposes of furthering knowledge and advising the federal government. Functioning in accordance with general policies determined by the Academy, the Council has become the principal operating agency of both the National Academy of Sciences and the National Academy of Engineering in providing services to the government, the public, and the scientific and engineering communities. The Council is administered jointly by both Academies and the Institute of Medicine. Dr. Frank Press and Dr. Robert M. White are chairman and vice chairman, respectively, of the National Research Council.

The Board on Science and Technology for International Development (BOSTID) of the Office of International Affairs addresses a range of issues arising from the ways in which science and technology in developing countries can stimulate and complement the complex processes of social and economic development. It oversees a broad program of bilateral workshops with scientific organizations in developing countries and conducts special studies. BOSTID's Advisory Committee on Technology Innovation publishes topical reviews of technical processes and biological resources of potential importance to developing countries.

This report has been prepared by an ad hoc advisory panel of the Advisory Committee on Technology Innovation, Board on Science and Technology for International Development, Office of International Affairs, National Research Council. Staff support was funded by the Office of the Science Advisor, Agency for International Development, under Grant No. DAN-5538-G-00-1023-00, Amendments 27 and 29.

Library of Congress Catalog Card Number: 91-68332
ISBN 0-309-04686-6
S527

Second printing 1993

PANEL ON NEEM

EUGENE B. SHULTZ, JR., School of Engineering and Applied Science, Washington University, St. Louis, Missouri, USA, *Chairman*

DEEPAK BHATNAGAR, Agricultural Research Service, U.S. Department of Agriculture, New Orleans, Louisiana, USA

MARTIN JACOBSON, U.S. Department of Agriculture (retired), Silver Spring, Maryland, USA

ROBERT L. METCALF, Department of Entomology, University of Illinois, Urbana, Illinois, USA

RAMESH C. SAXENA, International Centre of Insect Physiology and Ecology, Nairobi, Kenya

DAVID UNANDER, Division of Population Science, Fox Chase Cancer Center, Philadelphia, Pennsylvania, USA

* * *

NOEL D. VIETMEYER, Senior Program Officer, Board on Science and Technology for International Development, *Neem Study Director* and *Scientific Editor*

STAFF

F. R. RUSKIN, *BOSTID Editor*
ELIZABETH MOUZON, *Senior Secretary*
BRENT SIMPSON, *MUCIA Intern*

JOHN HURLEY, *Director* (until November 1991)

CONTRIBUTORS

SALEEM AHMED, East-West Center, Honolulu, Hawaii, USA
K.R.S. ASCHER, Department of Toxicology, The Volcani Center, Bet Dagan, Israel
EDWARD S. AYENSU, Pan-African Union for Science and Technology, Accra, Ghana
MICHAEL D. BENGE, U.S. Agency for International Development, Washington, D.C., USA
BARUCH S. BLUMBERG, Division of Population Science, Fox Chase Cancer Center, Philadelphia, Pennsylvania, USA
JEAN GORSE, Paris, France
JEFFREY GRITZNER, Public Policy Research Institute, University of Montana, Missoula, Montana, USA
BRUCE HARRISON, Board on Science and Technology for International Development, Washington, D.C., USA
MURRAY B. ISMAN, Department of Plant Science, University of British Columbia, Vancouver, British Columbia, Canada
C.M. KETKAR, Neem Mission, Maharashtra, India
T.N. KHOSHOO, Tata Energy Research Institute, New Delhi, India
JIM KLOCKE, ISK Mountain View Research Center, Sunnyvale, California, USA (deceased)
HIRAM LAREW, U.S. Agency for International Development, Washington, D.C., USA
ROBERT O. LARSON, Vikwood Botanicals, Inc., Sheboygan, Wisconsin, USA
DAVID PLUYMERS, College of Engineering and Applied Science, Washington University, St. Louis, Missouri, USA
MARTIN PRICE, ECHO, North Fort Myers, Florida, USA
STANISLAW RADWANSKI, Paris, France
HEINZ REMBOLD, Max-Planck-Institut für Biochemie, D-8033 Martinsried bei München, Germany
HEINRICH SCHMUTTERER, Institut für Phytopathologie und Angewandte Zoologie, Justus-Liebig-Universität, Giessen, Germany
PETER P. STRZOK, Agency to Facilitate the Growth of Rural Organizations, Minneapolis, Minnesota, USA
JAMES F. WALTER, W.R. Grace and Company-Conn., Columbia, Maryland, USA
DAVID WARTHEN, Agricultural Research Service, U.S. Department of Agriculture, Beltsville, Maryland, USA
GERALD E. WICKENS, Hampton Hill, Middlesex, England

Preface

Neem is a fascinating tree. On the one hand, it seems to be one of the most promising of all plants and may eventually benefit every person on the planet. Probably no other yields as many strange and varied products or has as many exploitable by-products. Indeed, as foreseen by some scientists, this plant may usher in a new era in pest control, provide millions with inexpensive medicines, cut down the rate of human population growth, and perhaps even reduce erosion, deforestation, and the excessive temperature of an overheated globe.

On the other hand, that all remains only a vague promise. Although the enthusiasm may be justified, it is largely founded on empirical or anecdotal evidence. Our purpose here is to marshal the various facts about this little-known species, to help illuminate its future promise, and to speed realization of its potential.

The report has been produced particularly for nonspecialists such as government ministers, research directors, university students, private voluntary organizations, and entrepreneurs. It is intended as an economic development document, not a scientific monograph. We hope it will be of interest, especially to agencies engaged in development assistance and food relief; officials and institutions concerned with agriculture and forestry in tropical countries; and scientific establishments with relevant interests.

This study is a project of the Board on Science and Technology for International Development (BOSTID), a division of the National Research Council. It is one in a series of reports prepared under BOSTID's program on technology innovation. Established in 1970, this program evaluates unconventional scientific and technological advances with particular promise for solving problems of developing countries. This report continues a subseries of reports describing promising tree resources that heretofore have been neglected or overlooked. Other titles include:

- *Leucaena: Promising Forage and Tree Crop in Developing Countries* (1984)
- *Mangium and Other Fast-Growing Acacias for the Humid Tropics* (1983)

- *Calliandra: A Versatile Small Tree for the Humid Tropics* (1983)
- *Casuarinas: Nitrogen-Fixing Trees for Adverse Sites* (1983)
- *Firewood Crops: Shrub and Tree Species for Energy Production*, Volumes I and II (1980 and 1983, respectively)
- *Sowing Forests from the Air* (1981).

Funds for this project were made available by the Agency for International Development (AID). Specifically, they were contributed by AID's Office of Forestry, Environment, and Natural Resources.

How to cite this report:
National Research Council. 1992. *Neem: A Tree For Solving Global Problems*. National Academy Press, Washington, D.C.

Contents

Art Credits

Page	
9	Reproduced with permission from *1988 Focus on Phytochemical Pesticides: Volume 1, the Neem Tree*, M. Jacobson, ed. CRC Press, Inc., Boca Raton, Florida, USA.
24	Peggy K. Duke
Cover Design	David Bennett

Foreword

The people of India have long revered the neem tree (*Azadirachta indica*). For centuries, millions have cleaned their teeth with neem twigs, smeared skin disorders with neem-leaf juice, taken neem tea as a tonic, and placed neem leaves in their beds, books, grain bins, cupboards, and closets to keep away troublesome bugs. The tree has relieved so many different pains, fevers, infections, and other complaints that it has been called "the village pharmacy."

To those millions in India neem has miraculous powers, and now scientists around the world are beginning to think they may be right. Two decades of research have revealed promising results in so many disciplines that this obscure species may be of enormous benefit to countries both poor and rich. Even some of the most cautious researchers are saying that "neem deserves to be called a wonder plant."

In particular, neem may be the harbinger of a new generation of "soft" pesticides that will allow people to protect crops in benign ways.

Although apparently justified by the evidence, the rising enthusiasm is based largely on exploratory investigations rather than controlled experiments or the widespread use of neem products in modern practice. The results have seldom, if ever, been subjected to the rigors of independent evaluation or use. Once that happens, everything may change.

Despite all the uncertainties, however, the possibilities are indeed intriguing. The following chapter, a composite of the visions of various researchers involved with neem, shows why.

Noel Vietmeyer
Study Director

1

The Vision

Native to India and Burma, neem is a botanical cousin of mahogany. It is tall and spreading like an oak and bears masses of honey-scented white flowers like a locust. Its complex foliage resembles that of walnut or ash, and its swollen fruits look much like olives. It is seldom leafless, and the shade it imparts throughout the year is a major reason why it is prized in India. The Subcontinent contains an estimated 18 million neem trees, most of them lined along roadsides or clustered around markets or backyards to provide relief from the sun.

Under normal circumstances neem's seeds are viable for only a few weeks, but earlier this century people somehow managed to introduce this Indian tree to West Africa, where it has since grown well. They probably expected neem to be useful only as a source of shade and medicinals—especially for malaria—but in Ghana it has become the leading producer of firewood for the densely populated Accra Plains, and in countries from Somalia to Mauritania it is a leading candidate for helping halt the southward spread of the Sahara Desert.

This century, people took neem seed to other parts of the world, where the tree has also performed well. Near Mecca, for example, a Saudi philanthropist planted a forest of 50,000 neems to shade and comfort the two million pilgrims who camp each year on the Plains of Arafat (a holy place where the prophet Muhammad is said to have bidden farewell to his followers). And in the last decade neem has been introduced into the Caribbean, where it is being used to help reforest several nations. Neem is already a major tree species in Haiti, for instance.

But neem is far more than a tough tree that grows vigorously in difficult sites. Among its many benefits, the one that is most unusual and immediately practical is the control of farm and household pests. Some entomologists now conclude that neem has such remarkable powers for controlling insects that it will usher in a new era in safe,

Miami, Florida. The neem tree is attractive, distinctive, and can grow to large size. Apart from masses of fruits and seeds, a big tree like this provides a dense shade that is much appreciated in the hot zones where it thrives. (Agricultural Research Service, USDA)

natural pesticides.[1] Extracts from its extremely bitter seeds and leaves may, in fact, be the ideal insecticides: they attack many pestiferous species; they seem to leave people, animals, and beneficial insects unharmed; they are biodegradable; and they appear unlikely to quickly lose their potency to a buildup of genetic resistance in the pests. All in all, neem seems likely to provide nontoxic and long-lived replacements for some of today's most suspect synthetic pesticides.

That neem can foil certain insect pests is not news to Asians. For centuries, India's farmers have known that the trees withstand the periodic infestations of locusts. Indian scientists took up neem research as far back as the 1920s, but their work was little appreciated elsewhere until 1959 when a German entomologist witnessed a locust plague in the Sudan. During this onslaught of billions of winged marauders, Heinrich Schmutterer noticed that neem trees were the only green things left standing. On closer investigation, he saw that although the locusts settled on the trees in swarms, they always left without feeding. To find out why, he and his students have studied the components of neem ever since.

Schmutterer's work (as well as a 1962 article by three Indian scientists showing that neem extracts applied to vegetable crops would repel locusts) spawned a growing amount of lively research. This, in turn, led to three international neem conferences, several neem workshops and symposia, a neem newsletter, and rising enthusiasm in the scientific community. By 1991, several hundred researchers in at least a dozen countries were studying various aspects of neem and its products.

Like most plants, neem deploys internal chemical defenses to protect itself against leaf-chewing insects. Its chemical weapons are extraordinary, however. In tests over the last decade, entomologists have found that neem materials can affect more than 200 insect species as well as some mites, nematodes, fungi, bacteria, and even a few viruses. The tests have included several dozen serious farm and household pests—Mexican bean beetles, Colorado potato beetles, locusts, grasshoppers, tobacco budworms, and six species of cockroaches, for example. Success has also been reported on cotton and tobacco pests in India, Israel, and the United States; on cabbage pests in Togo, Dominican Republic, and Mauritius; on rice pests in the Philippines; and on coffee bugs in Kenya. And it is not just the living

---

[1] In this report we use the words "pesticide" and "insecticide" in the broad sense of pest- and insect-controlling agents. By strict definition, the words imply toxins that kill outright; neem compounds, however, usually leave the pests alive for some time, but so repelled, debilitated, or hormonally disrupted that crops, people, and animals are protected.

plants that are shielded. Neem products have protected stored corn, sorghum, beans, and other foods against pests for up to 10 months in some very sophisticated controlled experiments and field trials.

Researchers at the U.S. Department of Agriculture have been studying neem since 1972. In laboratory experiments, they have found that the plant's ingredients foil even some of America's most voracious garden pests. For instance, in one trial each half of several soybean leaves was sprayed with neem extracts and placed in a container with Japanese beetles. The treated halves remained untouched, but within 48 hours the other halves were consumed right down to their woody veins. In fact, the Japanese beetles died rather than eat even tiny amounts of neem-treated leaf tissue. In field tests, neem materials have yielded similarly promising results. For instance, in one test in Ohio, soybeans sprayed with neem extract stayed untouched for up to 14 days; untreated plants in the same field were chewed to pieces by various species of insects, seemingly overnight.

Neem contains several active ingredients, and they act in different ways under different circumstances. These compounds bear no resemblance to the chemicals in today's synthetic insecticides. Chemically, they are distant relatives of steroidal compounds, which include cortisone, birth-control pills, and many valuable pharmaceuticals. Composed only of carbon, hydrogen, and oxygen, they have no atoms of chlorine, phosphorus, sulfur, or nitrogen (such as are commonly found in synthetic pesticides). Their mode of action is thus also quite different.

Neem products are unique in that (at least for most insects) they are not outright killers. Instead, they alter an insect's behavior or life processes in ways that can be extremely subtle. Eventually, however, the insect can no longer feed or breed or metamorphose, and can cause no further damage.

For example, one outstanding neem component, azadirachtin, disrupts the metamorphosis of insect larvae. By inhibiting molting, it keeps the larvae from developing into pupae, and they die without producing a new generation. In addition, azadirachtin is frequently so repugnant to insects that scores of different leaf-chewing species—even ones that normally strip everything living from plants—will starve to death rather than touch plants that carry traces of it.

Another neem substance, salannin, is a similarly powerful repellent. It also stops many insects from touching even the plants they normally find most delectable. Indeed, it deters certain biting insects more effectively than the synthetic chemical called "DEET" (N,N-diethyl-*m*-toluamide), which is now found in hundreds of consumer insect repellents.

To obtain the insecticides from this tree is simple (at least in

Although all parts of a neem tree are useful in various ways, it is the almondlike seed kernels (left) that have attracted the most scientific attention in recent years. Ground up and extracted with water or various solvents, these kernels can provide a range of potent pesticides, as well as neem oil (right) that has several industrial uses and may also be useful as a pesticide. (R. Nowitz, Agricultural Research Service, USDA)

principle). The leaves or seeds are merely crushed and steeped in water, alcohol, or other solvents. For some purposes, the resulting extracts can be used without further refinement. These pesticidal "cocktails," containing 4 major and perhaps 20 minor active compounds, can be astonishingly effective. In concentrations of less than one-tenth of a part per million, they affect certain insects dramatically. In trials in The Gambia, for example, these crude neem extracts compared favorably with the synthetic insecticide malathion in their effects on some of the pests of vegetable crops. In Nigeria, they equaled the effectiveness of DDT, Dieldrin, and other insecticides. And elsewhere in the world these plant products have often showed results as good as those of standard pesticides.

The extracts from neem seeds can also be purified and the most effective ingredients isolated from the rest of the mix. This process allows standardization and uniform formulations that can be produced for commercial use in even the world's most sophisticated pesticide markets.

Whatever the mixture or formulation, neem-based products display several remarkable qualities. For example, although pests can become tolerant to a single toxic chemical such as malathion, it seems unlikely

The extracts from neem kernels have surprising effects on many types of insects. Even tiny amounts disrupt the life cycles of more than 200 pest species. On these pages we show a few examples. Clockwise from top left:

- Desert locust. After four locust nymphs (fifth instar) were treated with neem oil, three got only halfway out of terminal molt and died. The fourth molted completely but had fatal deformities. (H. Schmutterer)
- Colorado potato beetle. These neem-treated specimens have malformed wings that can neither function in flight nor fold flat. In addition, these specimens are sterile. (E.B. Radcliffe)
- Grasshopper. After treating a third-instar nymph of the variegated grasshopper, *Zonocerus variegatus*, the adult emerged with no antennae and defective wings. (H. Schmutterer)
- Gypsy moth. Applying neem extracts to the larva (third instar) resulted in an incomplete molt. The old skin is still attached to the back of the body, making the insect unable to function. (H. Schmutterer)

that they can develop genetic resistance to neem's complex blend of compounds—many functioning quite differently and on different parts of an insect's life cycle and physiology. Certainly, they won't do so quickly. Several experiments have failed to detect any signs of incipient resistance to the mixture. For example, even after being exposed to neem for 35 successive generations, diamondback moths remained as susceptible as they had been at the beginning.

Another valuable quality is that some neem compounds act as systemic agents in certain plant species. That is, they are absorbed by, and transported throughout, the plants. In such cases, aqueous neem extracts can merely be sprinkled on the soil. The ingredients are then absorbed by the roots, pass up through the stems, and perfuse the upper parts of the plant. In this way, crops become protected from within. In trials, the leaves and stems of wheat, barley, rice, sugarcane, tomatoes, cotton, and chrysanthemums have been protected from certain types of damaging insects for 10 weeks in this way.

Because systemic materials are inside the plant, they cannot be washed off by rain. Nor can they harm bees, predacious insects, and other organisms that do not chew plant tissue. Even new growth that occurs following the treatment may be protected. (In the case of conventional sprayed-on chemicals, on the other hand, new growth is usually vulnerable to insects.)

Perhaps the most important quality is that neem products appear to have little or no toxicity to warm-blooded animals. Birds and bats eat the sweet pulp of the fruits of neem trees without apparent ill effects. In fact, neem fruits are a main part of their diets in some locations, such as on Ghana's Accra Plains. When neem-seed extracts were brushed on the skins of rats, the animals' blood showed no abnormalities; indeed, the treated rats ate more food and gained more weight than the untreated ones.

This safety to mammals apparently extends to people. The deaths of a few young children in Malaysia in the 1980s have been linked to the doses of neem-seed oil forced on them by their parents. (Like the previous use of castor oil in the Western world, neem oil in Asia is considered a cure-all for some childhood illnesses.) However, other than this, no hazard has been documented under conditions of normal usage. For one thing, neem extracts show no mutagenicity in the Ames test, which detects potential carcinogens. For another, people in India have been adding neem leaves to their grain stores for centuries to keep weevils away. Thus, for many generations millions have been eating traces of neem on a daily basis.

Certain neem products may even benefit human health. The seeds and leaves contain compounds with demonstrated antiseptic, antiviral, and antifungal activity. There are also hints that neem has anti-

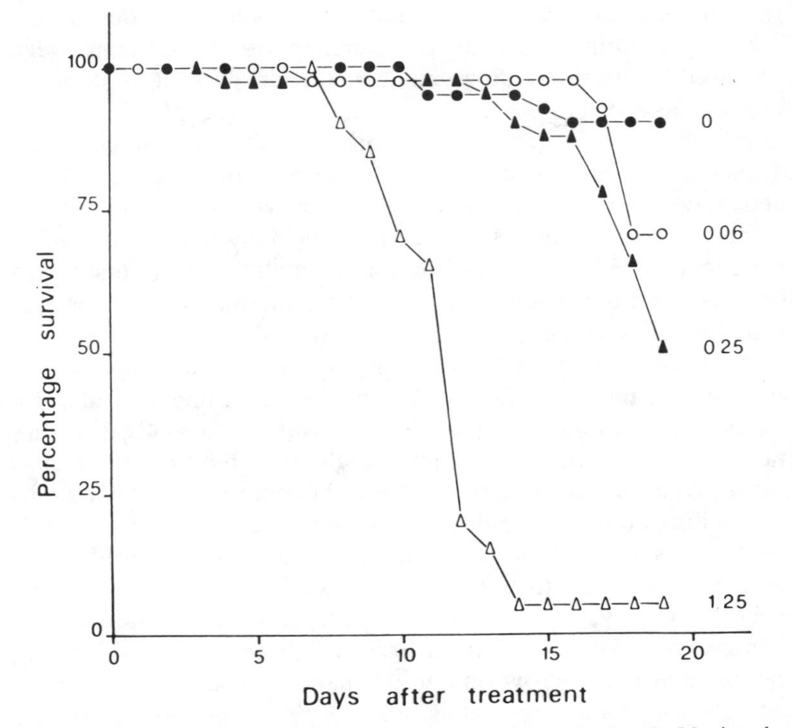

Deadly effect of azadirachtin. As shown in this laboratory assay using the Mexican bean beetle, neem products can control insect pests remarkably well. A concentration of little more than 1 part per million (ppm) resulted in essentially a complete kill within 2 weeks. With a concentration of merely one-fourth of a ppm, half of the insects were killed, but it took almost 3 weeks. In this test, the insects were raised on bean leaves in petri dishes. The methanolic solutions containing azadirachtin were distributed evenly over the leaves. This work was conducted by H. Rembold. (See Research Contacts, Appendix D).

inflammatory, hypotensive, and anti-ulcer effects. There is a potential indirect benefit to health as well. Neem leaves contain an ingredient that disrupts the fungi that produce aflatoxin on moldy peanuts, corn, and other foods—it leaves the fungi alive, but switches off their ability to produce aflatoxin, the most powerful carcinogen known.

For dental hygiene, especially, neem could prove valuable. Despite a general lack of toothpaste and toothbrushes, most people in India have bright, healthy teeth, and dental researchers usually attribute this to "chewsticks." Every morning, millions of Indians break off a twig, chew the end into a brushlike form, and scrub their teeth and gums. The most popular are the twigs from neem, and the selection seems to have a scientific basis. Research has shown, for example, that compounds in neem bark are strongly antiseptic. Also, tests in

Germany have proved that neem extracts prevent tooth decay, as well as both preventing and healing inflammations of the gums. Neem is now used as the active ingredient in certain popular toothpastes in Germany and India.

Moreover, researchers have recently found that neem might be able to play a part in controlling population growth. Materials from the seeds have been shown to have contraceptive properties. The oil is a strong spermicide and, when used intravaginally, has proven effective in reducing the birth rate in laboratory animals. A recent test involving the wives of more than 20 Indian Army personnel has further demonstrated its effectiveness.

Other neem compounds show early promise as the long-sought oral birth-control pill for men. This is just an intriguing hint at present; however, in exploratory trials they reduced fertility in male monkeys and a variety of other male mammals without inhibiting sperm production. In addition, the effects seemed to be temporary, which would be a big selling point that could help its rapid and widespread adoption.

All of this is potentially of vital importance for the world's poorest countries, many of which have high rates of population growth, severe problems with various agricultural pests, and a widespread lack of even basic medicine. The neem tree will grow in many Third World regions, and it can grow on certain marginal lands where it will not compete with food crops. Thus, it could bring good health and better crop yields within the reach of farmers too poor to buy pharmaceuticals or farm chemicals. It makes feasible the concept of producing one's own pesticide because the active materials can be extracted from the seeds, even at the farm or village level. Extracting the seeds requires no special skills or sophisticated machinery, and the resulting products can be applied using low-technology methods.

This possibility is significant because most developing countries are in the tropics, where year-round warmth often allows pest populations to build to unacceptable levels. The problems attendant on using synthetic pesticides, therefore, are particularly severe in the Third World. For instance, the World Health Organization attributes 20,000 deaths and more than a million illnesses each year to pesticides mishandled or used to excess. In addition, because the pests breed year-round, mutational resistance builds up much more quickly in the tropics than in lands having winter seasons.

Neem also seems particularly appropriate for developing country use because it is a perennial and requires little maintenance. It appeals to people in both rural and urban areas because (unlike most trees) its leaves, fruits, seeds, and various other parts can be used in a multitude of ways. Moreover, it can grow quickly and easily and does not necessarily displace other crops.

In West Africa, farmers have taken to neem avidly and have helped spread it throughout the region. They like the tree because it makes excellent windbreaks and provides dense shade and a multitude of products. They like its termite-resistant wood for use in both construction and carpentry. They can sell its seed oil for use in lubricants, oil-burning lamps, and soapmaking. Moreover, the tree grows quickly, often thrives in poor soils, and neutralizes soil acidity with its alkaline leaves. Only recently, however, has anyone begun to use neem products in pest control in West Africa. (CARE Photo by Michael Ahearn)

Neem and the United States

In 1975 the U.S. Department of Agriculture research facility at Beltsville, Maryland, and 19 of its stations across the country embarked on a comprehensive program to study the pest-control properties of various plants. Several universities collaborated on this program; others worked independently. The research, which still continues at several locations, has demonstrated or verified the outstanding effects of neem extracts against numerous species of destructive insects and fungi. Of thousands of plant extracts tested, neem was by far the best.

In trials, crude alcohol extracts of neem seeds proved effective at very low concentrations against 60 species of insects, 45 of which are extremely damaging to American crops and stored products—causing billions of dollars of losses to the nation each year. They included sweet-potato whitefly (see page 94), serpentine leafminers (which attack vegetable and flower crops), gypsy moth (which causes millions of dollars of losses to homeowners and the forest industry), and several species of cockroach.

In 1985, the Environmental Protection Agency approved a commercial neem-based insecticide for certain nonfood uses. Called Margosan-O®, the product is available at present in limited quantities in 21 states, and the amount is growing quickly. It is registered for use against such pests as whiteflies on chrysanthemums, leafminers on birch trees, aphids on roses, chinch bugs on lawns, gypsy moths on shade trees, and thrips on gladiolus. So far, it is being used primarily in professional greenhouses.

As a result of all this work, neem is seen as the nation's leading candidate for providing a new generation of broad-spectrum pesticides. However, neem cannot yet be legally used against pests that occur on food crops. Despite neem's apparent lack of toxicity or environmental danger, getting the authorization to use it on food plants will take time and great expense, because federal agencies require exhaustive testing before approving any pesticide for this purpose.

Although neem is essentially unknown to the American public, some neem-based consumer products are appearing in shops across the nation. Imported neem soap and toothpaste, for example, are sold fairly widely in specialty stores.

Neem production and processing also provides employment and generates income in rural communities—perhaps a small, but nonetheless valuable, benefit in these days of mass flight to the cities in a desperate search for jobs. It could be a useful export as well; a ton of neem seed already sells at African ports (Dakar, for example) at more than twice the price of peanuts. On top of all that, neem by-products (the seedcake and leaves, in particular) actually may improve the local soils and help foster sustainable crop production.

Although neem's ability to promote health and its value as a safe pest control is still only in the realm of possibility, there is no doubt that neem trees can provide the poor and the landless with oil, feed, fertilizer, wood, and other essential resources. In its crude state the oil from the seeds can have a strong garlic odor, but even in that form it can be used for heating, lighting, or crude lubricating jobs. Refined, it loses its unpleasant smell and is used in soaps, cosmetics, disinfectants, and other industrial products.

Neem cake, a solid material left after the oil is pressed from the seeds, is also useful. Broadcast over farm fields, it provides organic matter as well as some fertility to the soil. More important, it controls several types of soil pests. It is, for example, notably effective against nematodes, those virulent microscopic worms that suck the life out of many crops. Cardamom farmers in southern India claim that neem cake is as effective as the best nematode-suppressing commercial products.

Because neem is a tree, its large-scale production promises to help alleviate several global environmental problems: deforestation, desertification, soil erosion, and perhaps even (if planted on a truly vast scale) global warming. Its extensive, deep roots seem to be remarkably effective at extracting nutrients from poor soils. These nutrients enter the topsoil as the leaves and twigs fall and decay. Thus, neem can help return to productive use some worn-out lands that are currently unsuited to crops. It is so good for this purpose that a 1968 United Nations report called a neem plantation in northern Nigeria "the greatest boon of the century" to the local inhabitants.

At a more basic level, the increasing scientific scrutiny of neem is providing biological insight into the way plants protect themselves against the multitude of plant eaters. It is a window on a battle in the continuing chemical warfare between plants and predators. And because it is part of this natural antagonism, neem is a promising candidate for use in the increasingly popular concept of integrated pest management. To employ neem in pest control is to take advantage of the plant kingdom's 400 million years of experience at trying to frustrate the animal kingdom.

For all its apparent promise, however, the research on neem and the development of its products are not receiving the massive support that might seem justified. Indeed, all the promise mentioned above is currently known to only a handful of entomologists, foresters, and pharmacologists—and, of course, to the traditional farmers of South Asia. Much of the enthusiasm and many of the claims are sure to be tempered as more insights are gained and more field operations are conducted. Nonetheless, improving pest control, bettering health, assisting reforestation, and perhaps checking overpopulation appear to be just some of the benefits if the world will now pay more attention to this benevolent tree.

Among many new developments in the 20 months since the first printing of this book, our attention has been caught by the following.

- Three new neem-based products—Azatin®, Turplex®, and Align®—have entered the U.S. insecticide market.* The U.S. Environmental Protection Agency (EPA) has approved Align® for use on food and feed crops.
- Margosan-O® is now registered in all 50 states, and the EPA has approved it for use on food crops. Two related neem formulations, BioNeem® for the consumer market and Benefit® for lawn and turf care, are also available.†
- A neem newsletter has begun publication in the United States.‡
- More than 70,000 neem trees have been planted in Florida, Puerto Rico, and Mexico (Yucatan and Baja California).
- Ground-up neem leaves have been reported successful at treating scabies, a serious skin disease. Of 824 cases, 98 percent showed complete cures within 3-15 days.§
- Medical researchers in India have developed a topical neem-based product that appears to boost the body's defense against infection at the location where it is applied. It is being tested notably for protecting women from vaginal infections (viruses, bacteria, fungi, yeast) and pregnancy.‖

* The manufacturer is AgriDyne Technologies, Inc. (see Research Contacts, page 121).
† The manufacturer is W.R. Grace (see Research Contacts, page 121).
‡ Published by The Neem Association (see Research Contacts, page 121).
§ Information via Martin Price (see Research Contacts, page 121). The mite that causes scabies also causes mange in livestock (donkeys, camels, llamas, for instance).
‖ This development is led by Shakti N. Upadhyay of the National Institute of Immunology, Indian Council of Medical Research, P.O. Box 4508, New Delhi 110 029, India.

2
The Reality

Although the possibilities seem almost endless, nothing about neem is yet definite. The scientists who are most enthusiastic over the plant and its potential admit that at this stage the evidence to support their expectations is tentative. Even within the world of pest control its eventual place is by no means clear.

The truth is that despite all its properties and promise, some impediments must be overcome and many uncertainties clarified before neem's potential can be fully realized. These obstacles are summarized in this chapter; more detail can be found in later chapters.

DISADVANTAGES

By and large, the limitations known today all seem surmountable. Indeed, they present exciting challenges to the scientific and economic-development communities. Solving them may well bring a major new resource and a means for benefiting much of the world.

Lack of Experience

The greatest impediment to neem's commercial development may simply be a general lack of credibility, or even awareness, of what it is and what it can do. Neither the public, the majority of pesticide manufacturers, nor the health-care community in industrial countries now appreciate the plant or its promise. This is due in part to a lack of experience, in part to a lack of industrial interest (caused notably by the difficulty of patenting natural products), and in part to a lack of laboratory data to substantiate the claims. One researcher has called the neem scene an "uncharted jungle" of miscellaneous assertions, disconnected details, and limitless possibilities.

Genetic Variability

Another difficulty is caused by the fact that many of the neem trees scattered around the world are (for all intents and purposes) genetically distinct. This means that conclusions drawn from one may not be exactly applicable to the others. Extracts from neighboring trees, for instance, may differ in their mixtures of ingredients.

There is no current evidence that this has caused any practical problems. Eventually, however, certain elite types will undoubtedly be selected and propagated.

Lack of Registration

In an era when many people are desperately seeking alternatives to synthetic pesticides, it is ironic that neem's very uniqueness is slowing its acceptance by regulatory authorities. Neem components incapacitate pests by repelling them, stopping them from feeding, or upsetting their growth—only indirectly by killing them. Its varying modes of action, its complex and synergistic mixture of ingredients, and its lack of standardization all raise barriers that trouble pesticide regulators.

Lack of Standards

Writing regulations to cover neem has been made even more difficult because no standard of potency has yet been developed. For consistency of composition, a mixed product from nature cannot compete with a single molecule from a laboratory. For instance, the mix of active ingredients may vary with the sample's age, the locality where it was grown, the genes of the tree it came from, or the method by which the sample was handled or shipped. Moreover, the analytical techniques are tricky and, for the moment at least, the various reports of neem's level of efficacy cannot all be trusted.

Although the tree is easy to grow, the specific horticultural and climatic conditions that maximize its potency are still unknown. Extracts from trees grown in different parts of the world currently show differing levels of activity, and the relative differences vary with the types of insects being tested. Sorting out just how genetics and environment—not to mention handling methods and insect species—influence neem's various ingredients is a knotty problem. Experience may eventually prove, for example, that the best-looking seeds from the fastest-growing trees on the most advantageous sites produce the poorest pesticides.

Conflicting Approaches

It is one of neem's strengths that its ingredients can be used in formulations from the crudest to the most sophisticated. On the one hand, in a remote Third World village farmers may take a sack of crushed neem kernels, dunk it (like a tea bag) in a barrel of water overnight, and use the resulting "neem tea" on their vegetable crops the next day. On the other hand, the isolation of individual neem ingredients is already being conducted in sophisticated factory settings in the United States. This produces highly purified and certifiably uniform products that are a world away from neem tea in a tub in Thailand.

Both approaches are valid, of course, but their needs, priorities, costs, and objectives are so vastly different that people working at the two extremes may appear (even to themselves) to be working at cross-purposes. To the uninitiated, the conflicting views can make the whole neem concept seem unreliable.

Economic Uncertainties

The commercial production of any materials derived from the fruit of a tree is necessarily constrained by nature. There are limitations of seasonal supply, the long wait for the trees to mature, and the difficulty of facing the whims of nature. (For instance, in India neem fruits drop during the late monsoon, a time when frequent rains make them hard to dry.)

On the other hand, in many tropical nations neem pesticides should prove to be much cheaper than synthetic pesticides, and they could be homegrown rather than imported. However, they will never be totally without cost. Gathering and processing neem products takes time and effort. Even people "growing their own" will probably have to take time off from farming, fuel gathering, or other vital activities to harvest their neem seed. Moreover, if pests arrive during a season when fresh seeds are unavailable, facilities for storing the seeds for future use will be needed.

Contributing notably to the expense (at least in new plantings) is the delay of several years before the first crop can be gathered. Not only must the growers carefully nurture the young tree during its first year or so, they will begin getting returns only after its fifth year under normal conditions.[1]

---

[1] Not everyone will have to wait that long. For example, neem trees in the Dominican Republic have begun yielding fruit after just two years. (Information from H. Schmutterer.)

There can be other economic uncertainties as well. As we have noted, for instance, it is not yet known how best to manage the tree to optimize its production of pesticidal ingredients. Nor is it yet known if and how the mix or quality of the pest-control compounds will change in any economically meaningful way with the location, the climate, or the tree's age.

Handling Difficulties

Although neem products (seeds, extracts, or seedcake) are safe and easy to handle and apply, they are bulky and some samples smell like a dreadful cross between garlic and peanut butter.

There is at this point no method for mechanizing the process of collecting, storing, or handling the seeds. Nor is it yet known how to carry out these operations so that the pesticidal ingredients retain their fullest potency.

Geographical Limitations

Neem trees cannot be grown just anywhere. They are sensitive to frost and can be produced only in the warmer parts of the world. There is also mounting evidence that under dry conditions their growth and yield can be erratic and their susceptibility to pests high.

Planting Difficulties

The seed's short viability is a problem in introducing neem to new locations. Fresh seeds germinate well, but within weeks germination rates begin dropping off. This poses logistical difficulties for any tree-planting endeavors outside the areas where the tree now grows.

Silviculture Difficulties

At this stage at least, neem seems primarily suited for individual or group plantings within household compounds and villages, along roadsides and canals, in marketplaces and parks, and around the edges of fields. However, its production in organized commercial plantations in the long term might prove to be its greatest value.

Usually there is little difficulty with livestock eating the seedlings or saplings, but humans filching the foliage for medicinals or toothbrushes can be a problem.

Instability

When exposed to sunlight, neem products degrade and lose their pest-control properties. Typically, the crude extracts remain active for only eight days when exposed to the sun's ultraviolet rays.

Under sophisticated conditions this limitation can be overcome. The neem formulation being sold in the United States, for instance, contains sunscreens. When sprayed on plants, it remains effective for 2–3 weeks, and it can be stored for at least 2 years with little or no loss of potency.

Neem materials are also sensitive to high temperatures and must be stored in shady places.

These inherent instabilities can be exacerbated when the extracts are made under uncontrolled conditions, such as in a Third World village. For example, the active compounds may be inactivated by acids, alkalis, or other contaminants of local water supplies. Other types of pesticides might well be similarly affected, but neem extracts are more likely to be prepared where water is impure.

Health Hazards

Although neem has shown every indication of being safe to mammals in normal use as a pesticide (see Appendix A), the possibility of future hazards should not be dismissed. Few toxicity tests on higher mammals such as dogs, pigs, primates, or people have yet been published. As a result, in the United States neem products are not yet authorized for use on food crops. Their persistence in residues on foods is also unknown.

A known health hazard may arise as a result of poor handling. The harvested fruits must be depulped quickly and the seeds dried under shade and stored under shelter from the sun and rain. This is because at moisture contents above 14 percent, the fruits can carry the fungus *Aspergillus flavus*, which under many conditions produces aflatoxins. These are among the most potent carcinogens known and, unfortunately, they can contaminate the seeds inside the fruits. Indeed, they are extracted and concentrated along with the pesticidal ingredients. This may eventually prove to be the greatest barrier to the wider use of the pesticides from this most promising tree. It is, however, a problem only in the more humid neem-growing areas. Elsewhere, the climate is usually too dry for fungi to infect the fruits.

Environmental Safety

The fact that neem extracts are natural products does not mean that they are benign. Indeed, there is evidence that they can affect certain aquatic life. Most studies with fish in laboratory tests have shown no deleterious effects, but in one trial both tadpoles and the mosquito-eating fish gambusia died when neem extracts were applied to the water.[2] And neem seeds falling into fish ponds in Haiti killed tilapia

[2] Jotwani and Srivastava, 1981.

fry.[3] These experiences do not necessarily indicate an environmental hazard—only that caution and more toxicological studies are needed. It seems likely that neem oil, rather than the other seed-kernel ingredients, is causing the toxicity.

Although using neem will seldom harm beneficial insects, there are a few cases of negative effects. There is, for instance, a report of it affecting the larvae of hover flies. Also, there may be other subtle secondary effects. Bees and butterflies drinking nectar from neem-dosed plants might, for example, pick up traces of neem components, leading to reduced reproduction. The same may be said for insects that feed on other insects. To date, however, no evidence for such deleterious effects has surfaced.

Slow Action

Compared to DDT and other synthetic pesticides, the wait for neem to act may seem endless. Insects treated with it die by delayed action. Although their destructive power drops fast as the neem materials take effect, they may continue living for 2 weeks. Eventually, however, the next generation fails to emerge and the population collapses.

Although the end result may be more devastating than that from DDT, people used to seeing rapid knockdown may be initially disappointed, or even discouraged. This lack of quick effect poses a challenge for promoting neem in pest-control markets where people have come to expect instantaneous results.

Damage to Plants

It is one of neem's most exciting features that its compounds are systemic. However, they are not systemic in all plant species. Potato plants, for example, do not take up the main active ingredient, azadirachtin, whereas beans do. This introduces yet another uncertainty. Each plant species may have to be checked individually. Also, the acidity of the soil or the level of enzymatic activity in the plant may affect the length of time that neem compounds remain effective inside the plant tissues.

Moreover, it has been found in greenhouse and field trials that certain neem materials can damage plants. In cabbages, for example, only medium-sized heads were formed. In onions, the waxy coating on the leaves was destroyed. In tomatoes, the growth and yield were reduced.

Much of this "phytotoxicity" was apparently due to neem oil contaminating the samples. There were large differences between the

[3] Information from P. Welle.

damage caused by crude fractions and by refined products. It could be, therefore, that only highly purified extracts can be reliably used for systemic purposes. At normal levels, these have so far proved safe to even sensitive plants.

Method of Application

Although neem products can be applied using standard equipment (both sophisticated and primitive), they may have specific requirements if they are to be fully effective. For example, some pests must be treated at a certain time of day. Colorado potato beetle is one. If potato fields are sprayed when the sun is high, the extracts dry out and have little effect. On the other hand, if sprayed at dawn, they can be extremely effective.

A product like this, which affects certain subtle aspects of an insect's life, is restricted by factors having to do with the insect's habits, life stage, and metabolic processes. In many cases the users will have to be educated, or at least trained, before neem can be fully effective.

Protecting the Tree

Despite the fact that it is a source of pesticidal materials, the tree itself is attacked by certain pests. In 1986, for example, an outbreak of the oriental yellow scale was confirmed in West Africa (see next chapter). This insect, a native of India and the Far East, defoliates and sometimes kills the tree. Recent reports suggest that it has severely damaged neem trees over large areas of northern Cameroon, Chad, northeastern Nigeria, and eastern Niger. It seems likely that this outbreak resulted from the stresses of a decade or more of drought in the Sahel, which has left many neems weak and sickly.[4] Nonetheless, further devastating pest outbreaks are certainly possible.

Medical Limitations

Whereas millions of Indians swear to the efficacy of neem treatments, the pharmacological effects have seldom been subjected to rigorous trials with controls. Thus, to officials in many other parts of the world, today's claims of medical efficacy are suspect. Indeed, a couple of recent studies suggest that it may be unsafe to eat neem products.

One study (already mentioned) dealt with the use of neem oil as a general cure-all for children. It strongly suggests that this is a most unwise practice, at least among the very young. When children under

[4] Although the neems were weakened, many other tree species (*Acacia albida*, for example) were killed outright. In some places, the water table dropped by 20 m or more during the decade of drought.

the age of four were given doses (5–30 ml) of neem oil, they came down with a disease similar to Reye's syndrome.[5] This severe disorder involves swelling of the brain, liver, and other organs. Both Reye's syndrome and its neem-oil-induced mimic are poorly understood. Nonetheless, the message is clear: neem oil and neem extracts should not be used for internal medicinal purposes until more thoroughly tested.

There are also anecdotal accounts from West Africa of neem-leaf teas possibly causing kidney damage when taken over a long period.

In other studies, neem extracts were found to be toxic to guinea pigs and rabbits,[6] and leaves fed to goats and guinea pigs (50–200 mg per kg body weight) reduced their rate of growth.[7]

The ultimate significance of these preliminary studies is unclear. The materials used may well have been contaminated—perhaps by fungal toxins. Other trials have found no toxicity problem. For instance, in a toxicological study in Germany, neem oil obtained from clean, fungus-free seed kernels showed no oral toxicity in rats. The dosage tested was 5,000 mg per kg body weight.[8]

Certainly, no hazard has been observed when neem has been used in topical treatments (on skin complaints, for example) or in dental uses—which together make up by far the major medical applications. Nor is there any evidence that using the seed-kernel extracts as pesticides is hazardous to health.

Whether neem oil is safe for possible use as an intravaginal spermicide is also unclear. However, in this use there are perhaps even greater uncertainties. Contraception is a topic of such sensitivity—personal, political, scientific, and religious—that here, too, its future role is uncertain.

Its use as a spermatocide—the male contraceptive mentioned earlier—is so uncertain that it will likely take decades of research to develop even if no safety hazards are found along the way.

Synthetic Competition

The number and complexity of the compounds in neem extracts will always preclude the economic synthesis of the full mixture. On the other hand, individual compounds may prove suitable for synthesis. There is, therefore, the possibility that if neem opens up a new generation of pesticides, synthetic mimics may capture some of the more lucrative "top-end" markets.

---

[5] Sinniah et al., 1982; Sinniah et al., 1985. Reye's syndrome occurs primarily in children after viral illnesses and is associated with aspirin usage.
[6] Sadre et al., 1984.
[7] Ali, 1987.
[8] Information from H. Schmutterer.

3

The Tree

Neem is a member of the mahogany family, Meliaceae. It is today known by the botanic name *Azadirachta indica* A. Juss. In the past, however, it has been known by several names, and some botanists formerly lumped it together with at least one of its relatives. The result is that the older literature is so confusing that it is sometimes impossible to determine just which species is being discussed.[1]

DESCRIPTION

Neem trees are attractive broad-leaved evergreens that can grow up to 30 m tall and 2.5 m in girth. Their spreading branches form rounded crowns as much as 20 m across. They remain in leaf except during extreme drought, when the leaves may fall off. The short, usually straight trunk has a moderately thick, strongly furrowed bark. The roots penetrate the soil deeply, at least where the site permits, and, particularly when injured, they produce suckers. This suckering tends to be especially prolific in dry localities.

Neem can take considerable abuse. For example, it easily withstands pollarding (repeated lopping at heights above about 1.5 m) and its topped trunk resprouts vigorously. It also freely coppices (repeated lopping at near-ground level). Regrowth from both pollarding and coppicing can be exceptionally fast because it is being served by a root system large enough to feed a full-grown tree.

The small, white, bisexual flowers are borne in axillary clusters. They have a honeylike scent and attract many bees. Neem honey is popular, and reportedly contains no trace of azadirachtin.

---

[1] Previous botanic names were *Melia indica* and *M. azadirachta*. The latter name (not to mention neem itself) has sometimes been confused with *M. azedarach*, a West Asian tree commonly known as Persian lilac, bakain, dharak, or chinaberry. The taxonomy of all these closely related species is so complex that some botanists have recognized as many as 15 species; others, as few as 2.

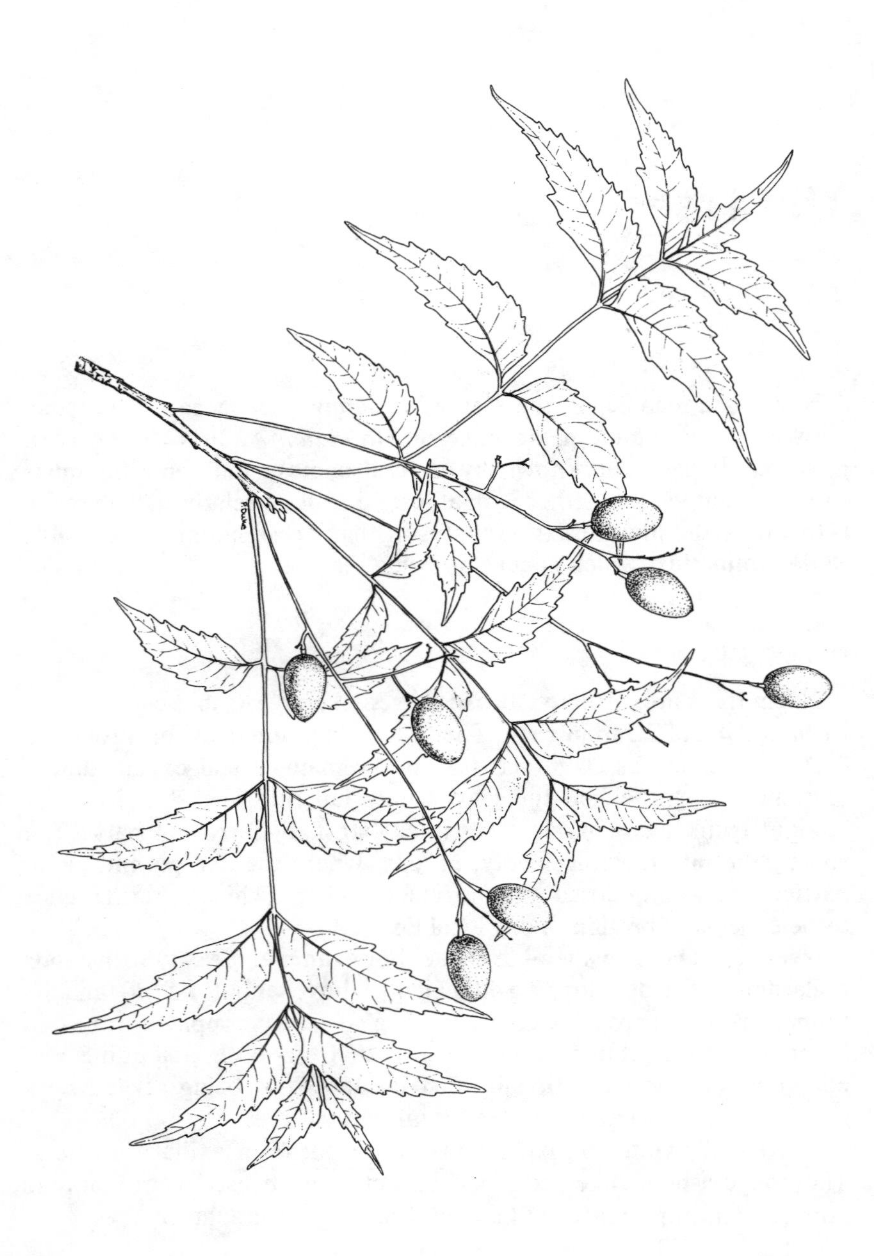

The fruit is a smooth, ellipsoidal drupe, up to almost 2 cm long. When ripe, it is yellow or greenish yellow and comprises a sweet pulp enclosing a seed. The seed is composed of a shell and a kernel (sometimes two or three kernels), each about half of the seed's weight. It is the kernel that is used most in pest control. (The leaves also contain pesticidal ingredients, but as a rule they are much less effective than those of the seed.)

A neem tree normally begins bearing fruit after 3–5 years, becomes fully productive in 10 years, and from then on can produce up to 50 kg of fruits annually. It may live for more than two centuries.

DISTRIBUTION

Neem is thought to have originated in Assam and Burma (where it is common throughout the central dry zone and the Siwalik hills). However, the exact origin is uncertain: some say neem is native to the whole Indian subcontinent; others attribute it to dry forest areas throughout all of South and Southeast Asia, including Pakistan, Sri Lanka, Thailand, Malaysia, and Indonesia.

It is in India that the tree is most widely used. It is grown from the southern tip of Kerala to the Himalayan hills, in tropical to subtropical regions, in semiarid to wet tropical regions, and from sea level to about 700 m elevation.

As already noted, neem was introduced to Africa earlier this century (see sidebar, page 85). It is now well established in at least 30 countries, particularly those in the regions along the Sahara's southern fringe, where it has become an important provider of both fuel and lumber. Although widely naturalized, it has nowhere become a pest. Indeed, it seems rather well "domesticated": it appears to thrive in villages and towns.

Over the last century or so, the tree has also been established in Fiji, Mauritius, the Caribbean, and many countries of Central and South America. In some cases it was probably introduced by indentured laborers, who remembered its value from their days of living in India's villages. In other cases it has been introduced by foresters. In the continental United States, small plantings are prospering in southern Florida, and exploratory plots have been established in southern California and Arizona.

PROPAGATION

The tree is easily propagated—both sexually and vegetatively. It can be planted using seeds, seedlings, saplings, root suckers, or tissue culture. However, it is normally grown from seed, either planted directly on the site or transplanted as seedlings from a nursery.

The seeds are fairly easy to prepare. The fruit drops from the trees by itself; the pulp, when wet, can be removed by rubbing against a coarse surface; and (after washing with water) the clean, white seeds are obtained. In certain nations—Togo and Senegal, for example—people leave the cleaning to the fruit bats and birds, who feed on the sweet pulp and then spit out the seeds under the trees.

It is reputed that neem seeds are not viable for long. It is generally considered that after 2–6 months in storage they will no longer germinate. However, some recent observations of seeds that had been stored in France indicated that seeds without endocarp had an acceptable germinative capacity (42 percent) after more than 5 years.[2]

GROWTH

The tree is said to grow "almost anywhere" in the lowland tropics. However, it generally performs best in areas with annual rainfalls of 400–1,200 mm. It thrives under the hottest conditions, where maximum shade temperature may soar past 50°C, but it will not withstand freezing or extended cold. It does well at elevations from sea level to perhaps 1,000 m near the equator. The taproot (at least in young specimens) may be as much as twice the height of the tree.

Neem is renowned for good growth on dry, infertile sites. It performs better than most trees where soils are sterile, stony, and shallow, or where there is a hardpan near the surface. The tree also grows well on some acid soils. Indeed, it is said that the fallen neem leaves, which are slightly alkaline (pH 8.2), are good for neutralizing acidity in the soil. On the other hand, neem cannot stand "wet feet," and quickly dies if the site becomes waterlogged.

Neem often grows rapidly. It can be cut for timber after just 5–7 years. Maximum yields reported from northern Nigeria (Samaru) amounted to 169 m^3 of fuelwood per hectare after a rotation of 8 years. Yields in Ghana were recorded between 108 and 137 m^3 per hectare in the same time.

Weeds seldom affect growth. Except in the case of very young plants, neem can dominate almost all competitors. In fact, the trees themselves may become "weeds." They spread widely under favorable site conditions, since the seeds are distributed by birds, bats, and baboons. For the same reason, natural regeneration under old trees is

[2] Information from Y. Roederer and R. Bellefontaine. Refrigeration is also said to extend the viability.

Neem fruits. The olivelike fruits can occur in large numbers on the tree. Animals like the fleshy outer part, but the seed in the center is the source of the ingredients people use most. (H. Schmutterer)

Common Names

English: neem, Indian lilac
French: azadira d'Inde, margousier, azidarac, azadira
Portuguese: margosa (Goa)
Spanish: margosa, nim
German: Niembaum
Hindi: neem, nimb
Burmese: tamar, tamarkha
Urdu: nim, neem
Punjabi: neem
Tamil: vembu, veppan
Sanskrit: nimba, nimbou, arishtha (reliever of sickness)
Sindi: nimmi
Sri Lanka: kohomba
Farsi: azad darakht i hindi (free tree of India), nib
Malay: veppa
Singapore: kohumba, nimba
Indonesia: mindi
Nigeria: dongoyaro
Kiswahili: mwarubaini (muarobaini)

often abundant. But for all that, in virtually every place it grows neem is considered a boon, not a bane. People almost always like to see more neems coming up.

PROBLEMS

Neem is renowned for its robust growth and resilience to harsh conditions, but, like all living things, it has various shortcomings, some of which are discussed below.

Pests

By and large, most neem trees are reputed to be remarkably pest free; however, in Nigeria 14 insect species and 1 parasitic plant have been recorded as pests.[3] Few of the attacks were serious, and the trees almost invariably recovered, although their growth and branching may have been affected.

[3] H. Schmutterer, 1990.

However, in recent years a more serious threat has emerged. In some parts of Africa (mainly in the Lake Chad Basin), a scale insect (*Aonidiella orientalis*) has become a serious pest. This and other scale insects sometimes infest neem trees in central and south India. They feed on sap, and although they do little harm to mature trees, they may kill young ones. Now that one type has been detected in Africa, the impact could be severe.[4]

Other insect pests include the following:

- The scale insect *Pinnaspis strachani* (very common in Asia, Africa, and Latin America);
- Leaf-cutting ants *Acromyrmex* spp. (common defoliators of young neem trees in Central and South America);
- The tortricid moth *Adoxophyes aurata* (attacks leaves in Asia including Papua New Guinea);
- The bug *Helopeltis theivora* (considered a serious neem pest in southern India); and
- The pyralid moth *Hypsipyla* sp. (attacks neem shoots in Australia).

Even though neem timber is renowned for termite resistance, termites sometimes damage, or even kill, the living trees. They usually attack only sickly specimens, however.

Diseases

Despite the fact that the leaves contain fungicidal and antibacterial ingredients, certain microbes may attack different parts of the tree, including the following:

- Roots (root rot, *Ganoderma lucidum*, for instance);
- Stems and twigs (the blight *Corticium salmonicolor*, for example);
- Leaves (a leaf spot, *Cercospora subsessilis*; powdery mildew, *Oidium* sp., and the bacterial blight *Pseudomonas azadirachtae*);[5] and
- Seedlings (several blights, rots, and wilts—including *Sclerotium*, *Rhizoctonia*, and *Fusarium*).[6]

A canker disease that discolors the wood and seems to coincide with a sudden absorption of water after long droughts has also been observed.[7]

---

[4] It has been suggested that a drastic lowering of the groundwater level around Lake Chad—which nearly dried out during a drought in the Sahel—was the main reason for the outbreak. This is perhaps true; scale insects are usually "secondary" pests that multiply best on plants that are already damaged by other pests or other adverse environmental factors.

[5] Bakshi, 1976; Desai et al., 1966.

[6] Sankaram et al., 1987.

[7] Information from J. Gorse.

Nutrient Deficiencies

A lack of zinc or potassium drastically reduces growth. Trees affected by zinc deficiency show chlorosis of the leaf tips and leaf margins, their shoots exude much resin, and their older leaves fall off. Those with potassium deficiency show leaf tip and marginal chlorosis and die back (necrosis).[8]

Other Problems

Fire kills neem seedlings outright. However, mature trees almost always regrow, especially if the dead parts are quickly cut away.

High winds are a potential problem. Large trees frequently snap off during hurricanes, cyclones, or typhoons. Neem is therefore a poor candidate for planting in areas prone to such violent storms.

Seedlings regenerating beneath stands of neem are sensitive to sudden exposure to intense sunlight. Thus, clear-felling neem trees normally produces a massive seedling kill, especially if the seedlings are small.[9]

In some localities rats and porcupines kill young trees by gnawing the bark around the base. Even when not causing any physical damage, rodents can be pests: wherever they are numerous, the fruits may disappear before the farmer can harvest them.

Neem, with its intensely bitter foliage, is not a preferred browse, but if nothing else is available goats and camels will eat it. In fact, in Asia goats and camels have been known to browse young neem trees so severely in times of scarcity that the plants died.[10] In Africa neem is generally ignored by livestock (which makes the tree easy to establish even within villages and courtyards). The reason that livestock treat neem differently in Asia and Africa is unknown at present. It may be differences in the tree specimens, or in the animals' preferences or past experiences.

[8] Zech, 1984.

[9] Information from R.W. Fishwick.

[10] Indeed, leaves could be used as a reserve fodder for camels, sheep, and goats. Despite their repellent bitterness, the leaves have a low fiber content and a high nutritional value (15 percent protein), comparable to that of leguminous leaves. However, there are also records of toxic effects of neem leaves on goats (Sudan).

4
What's in a Neem

Neem protects itself from the multitude of pests with a multitude of pesticidal ingredients. Its main chemical broadside is a mixture of 3 or 4 related compounds, and it backs these up with 20 or so others that are minor but nonetheless active in one way or another. In the main, these compounds belong to a general class of natural products called "triterpenes"; more specifically, "limonoids."

LIMONOIDS

So far, at least nine neem limonoids have demonstrated an ability to block insect growth, affecting a range of species that includes some of the most deadly pests of agriculture and human health. New limonoids are still being discovered in neem, but azadirachtin, salannin, meliantriol, and nimbin are the best known and, for now at least, seem to be the most significant.

Azadirachtin

One of the first active ingredients isolated from neem, azadirachtin has proved to be the tree's main agent for battling insects. It appears to cause some 90 percent of the effect on most pests. It does not kill insects—at least not immediately. Instead it both repels and disrupts their growth and reproduction. Research over the past 20 years has shown that it is one of the most potent growth regulators and feeding deterrents ever assayed. It will repel or reduce the feeding of many species of pest insects as well as some nematodes. In fact, it is so potent that a mere trace of its presence prevents some insects from even touching plants.

Azadirachtin is structurally similar to insect hormones called "ecdysones," which control the process of metamorphosis as the insects pass from larva to pupa to adult. It affects the corpus cardiacum, an

The Breakthrough

Although thousand-year-old Sanskrit medical writings mention neem's usefulness, the tree's exciting potential for controlling insects has only recently become clear.

Neem's ability to repel insects was first reported in the scientific literature in 1928 and 1929. Two Indian scientists, R.N. Chopra and M.A. Husain, used a 0.001-percent aqueous suspension of ground neem kernels to repel desert locusts. Not until 1962, however, was the real significance demonstrated. That year, in field tests in New Delhi, S. Pradhan ground up neem kernels in water and sprayed the resulting suspension over different crops. He found that, although locusts landed on the plants, they refused to eat anything, sometimes for up to 3 weeks after the treatment. Furthermore, he noted that neem kernels were even more potent than the conventional insecticides then available and that neem's repellency was as important as its toxicity. In neighboring insecticide-treated fields, for instance, the insects also died, but not before consuming the crops.

Neem's insect-growth-regulating (IGR) effects were independently observed in England and Kenya in 1972. In England, L.N.E. Ruscoe, at that time an employee of the ICI Company, tested azadirachtin on insect pests such as cabbage white butterfly (*Pieris brassicae*) and cotton stainer bug (*Dysdercus fasciatus*) and noted IGR effects in each case. The azadirachtin was provided by D. Morgan, a Keele University chemist who had been the first to isolate azadirachtin. In Kenya that same year, K. Leuschner, a German graduate student working at the Coffee Research Station in Upper Kiambu, observed that a methanolic neem-leaf extract controlled the coffee bug (*Antestiopsis orbitalis bechuana*) by growth-regulating effects. Most fifth-instar nymphs treated with the extract died during subsequent molts and the few that survived to adulthood had malformed wings and thoraxes.

Neem's fecundity-reducing effects were first recorded by R. Steets (another graduate student) and H. Schmutterer in Germany. Applying methanolic neem-kernel extract and azadirachtin to the Mexican bean beetle (*Epilachna varivestis*) and the Colorado potato beetle (*Leptinotarsa decemlineata*) they found that females almost stopped laying eggs. Some females had been completely sterilized, and the effect was irreversible.

organ similar to the human pituitary, which controls the secretion of hormones. Metamorphosis requires the careful synchrony of many hormones and other physiological changes to be successful, and azadirachtin seems to be an "ecdysone blocker." It blocks the insect's production and release of these vital hormones. Insects then will not molt. This of course breaks their life cycle.

On average, neem kernels contain between 2 and 4 mg of azadirachtin per gram of kernel. The highest figure so far reported—9 mg per g—was measured in samples from Senegal.

Meliantriol

Another feeding inhibitor, meliantriol, is able, in extremely low concentrations, to cause insects to cease eating. The demonstration of its ability to prevent locusts chewing on crops was the first scientific proof for neem's traditional use for insect control on India's crops.

Salannin

Yet a third triterpenoid isolated from neem is salannin. Studies indicate that this compound also powerfully inhibits feeding, but does not influence insect molts. The migratory locust, California red scale, striped cucumber beetle, houseflies, and the Japanese beetle have been strongly deterred in both laboratory and field tests.

Nimbin and Nimbidin

Two more neem components, nimbin and nimbidin, have been found to have antiviral activity. They affect potato virus X, vaccinia virus, and fowl pox virus. They could perhaps open a way to control these and other viral diseases of crops and livestock.

Nimbidin is the primary component of the bitter principles obtained when neem seeds are extracted with alcohol. It occurs in sizable quantities—about 2 percent of the kernel.

Others

Certain minor ingredients also work as antihormones. Research has shown that some of these minor neem chemicals even paralyze the "swallowing mechanism" and so prevent insects from eating. Examples of these newly found limonoids from neem include deacetylazadirachtinol. This ingredient, isolated from fresh fruits, appears to be as effective as azadirachtin in assays against the tobacco budworm, but it has not yet been widely tested in field practice.[1]

---

[1] Jacobson, 1986b.

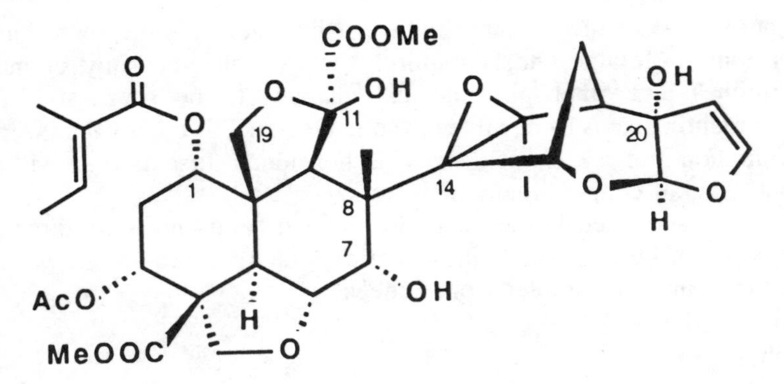

azadirachtin

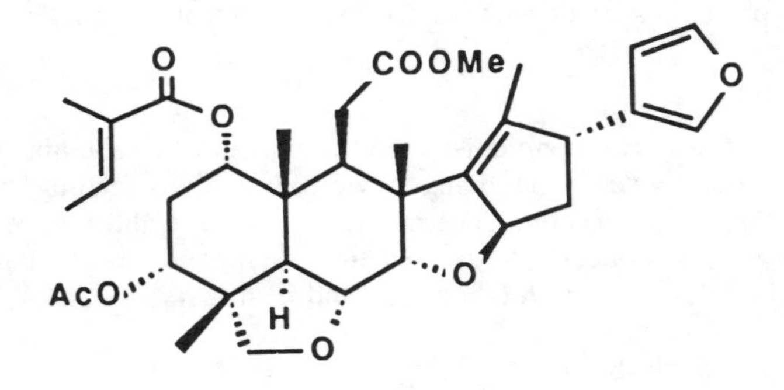

salannin

Chemical structures of neem's main ingredients. The complexity of these compounds demonstrates that nature is still the greatest chemist. Of the numerous pesticidal agents isolated so far from neem kernels, azadirachtin is the most active against insects. In addition to inhibiting their growth, it interferes with their powers of taste. Many leaf-eating insects are repelled by plants to which even small amounts of azadirachtin have been applied.

Azadirachtin, salannin, and nimbin all have the same basic limonoid structure. This differs from, but is not unlike, that of the sterols to which the insect molting hormones ("ecdysones") belong. An insect ingesting traces of these compounds is deeply affected because these "hormone mimics" block the parts of the brain that produce the hormones necessary to growth and development. In many cases, for instance, the insect's body may be ready to change while the hormones to complete the molt are not available. These deep-seated hormonal effects are the reason for neem's subtle, powerful, and yet insect-specific influences.

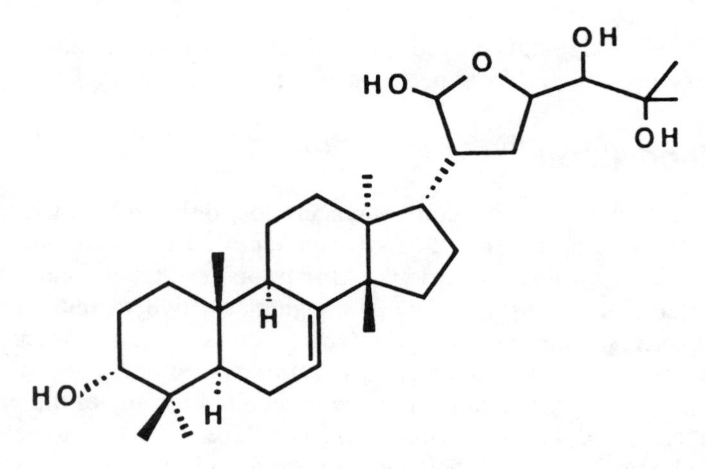

meliantriol

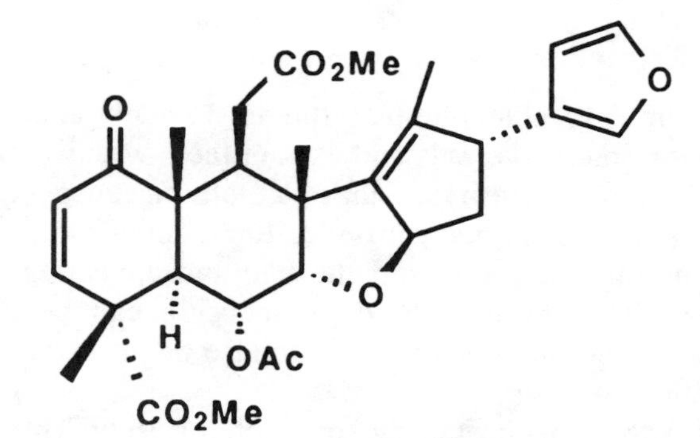

nimbin

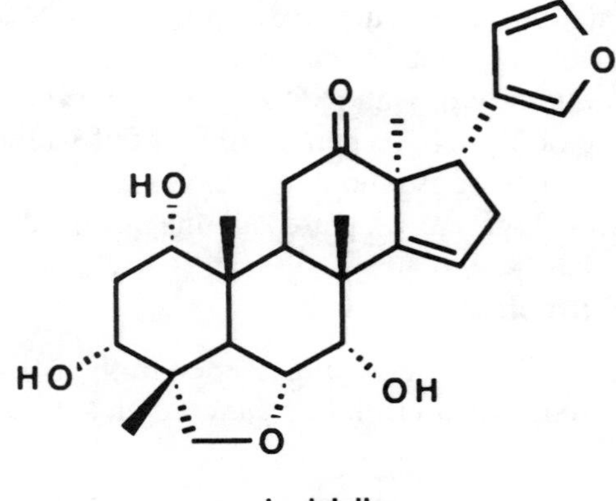

nimbidin

Two compounds related to salannin, 3-deacetylsalannin and salannol, recently isolated from neem, also act as antifeedants.

PRODUCTION

Although bioactive compounds are found throughout the tree, those in the seed kernels are the most concentrated and accessible. They are obtained by making various extracts of the kernels and, to a lesser extent, of the press cake. Although the active ingredients are only slightly soluble in water, they are freely soluble in organic solvents such as hydrocarbons, alcohols, ketones, or ethers.

No new or unusual technology is required for any of the processing. It can be done using either simple village-scale technology or high-technology methods and industrialized facilities. The most common procedures are summarized below.[2]

Water Extraction

The simplest technique (and the most widely employed today) is to crush or grind the kernels and extract them with water. They may, for example, be steeped overnight in a cloth bag suspended in a barrel of water. For reasons not yet understood, this process is less effective than pouring the water into the bag and collecting the extract as it emerges. The resulting crude suspension can be used in the field without further modification. It can also be filtered and employed as a sprayable emulsion.

This is the most promising approach for use in Third World villages. It has been estimated that by using water extraction, 20–30 kg of neem seed can normally treat 1 hectare. At this rate, the annual seed crop from one mature tree could treat up to half a hectare.[3] However, it is necessary to use a lot of water because the active ingredients have very low solubility in water. Normally, the proportions employed are about 500 g of kernel steeped in 10 liters of water.

Water extracts of ground neem leaves are also very useful. Because neem is an evergreen, they are obtainable throughout the year.

Hexane Extraction

If the kernels are grated and steeped in the solvent hexane, only the oil is removed. The oil is not considered an active pesticide. However,

[2] Preparations of aqueous, oil, powder, and press cake formulations, as well as other methods, are described in more detail in Stoll, 1986.

[3] Such figures are only a rough average. The amount used, of course, is different for each pest and planting density.

new results show that it is an especially interesting material, which in certain cases can be used to kill the eggs of many types of insect, the larvae of mosquitoes, and various stages of certain pests (such as leafhoppers) that are often hard to control by other means.

The residue left after the hexane extraction still contains the main active limonoid ingredients, and subsequent extractions with water or alcohol produce them in large amounts, clean and uncontaminated by oil.

Pentane Extraction

Pentane extracts of seed kernels are effective against spider mites. They reduce the fecundity (number of eggs) of *Tetranychus urticae*, for example. The active principles in the extracts differ from azadirachtin.[4]

Alcohol Extraction

Alcohol extraction is the most direct process for producing neem-based pesticidal materials in concentrated form. Limonoids are highly soluble in alcohol solvents. The grated kernels are usually soaked in ethanol, but sometimes in methanol. The yield of active ingredients varies from 0.2 to 6.2 percent.[5]

Although water extracts are effective as pesticides, neem compounds are not highly soluble in water; the alcohol extracts are about 50 times more concentrated. They may contain 3,000 parts per million (ppm) or even 100,000 ppm azadirachtin.

FORMULATIONS

As noted, the simplest neem pesticide is a crude extract. However, for more sophisticated use, various modifications can be made. These advanced formulations may convert neem extracts into the form of granules, dust, wettable powders, or emulsifiable concentrates. Aqueous extracts can also be formulated with soap for ease of application against skin diseases.

Other formulations may involve the addition of chemicals or even the chemical modification of the neem ingredients themselves. These changes may be made to increase shelf stability and reproducibility, and for ease of handling or of scaling up the process. They may also reduce phytotoxicity, the damage to sensitive plants.

---

[4] Sanguanpong and Schmutterer, 1991.

[5] Good rates of extraction have been obtained with other solvents as well. In comparing extraction methods, it was found that the azeotropic mixture of methanol and methyl tertiary-butyl ether was efficient and simple. It achieved an extract yield of 4–5 percent. (Feuerhake, 1984.)

One particularly valuable class of additives are those that inhibit ultraviolet degradation. These include sesame oil, lecithin, and *para*-aminobenzoic acid (PABA).

Additives

Mixing neem extracts with other materials can boost their power 10- to 20-fold. Among these so-called "promoters" are sesame oil, pyrethrins (a type of insecticide mostly extracted from chrysanthemum flowers, see sidebar page 91), and piperonyl butoxide. They are used to produce a quicker kill.

Combinations with synthetic pesticides also can work well—they add rapid "knockdown" to neem's ability to suppress the subsequent rebound in the pest population. The effectiveness of neem extracts can even be boosted with the insect-killing bacterium *Bacillus thuringensis* (Bt) to provide a multifaceted pesticide.

METHODS OF APPLICATION

Neem extracts can be applied in many ways, including some of the most sophisticated. For example, they may be employed as sprays, powders, drenches, or diluents in irrigation water—even through trickle- or subsurface-irrigation systems. In addition, they can be applied to plants through injection or topical application, either as dusts or sprays. Moreover, they can be added to baits that attract insects (a process used, for instance, with cockroaches). They are even burned. For example, neem leaves and seeds and dry neem cake are ingredients in some mosquito coils.

SYSTEMIC EFFECT

The fact that the extracts can be taken up by plants (and thereby confer protection from within) is one of neem's most interesting and potentially useful features. As has been noted, however, the level of this systemic activity differs from plant to plant and formulation to formulation. Extracts without oil, with a little oil, and with much oil exhibit different levels of systemic action.

The systemic activity differs with the insect as well. It is not effective on some aphids, for instance. They feed in phloem tissues, where (for reasons yet unknown) the concentration of azadirachtin is very low. Phloem is the plant's outermost layer of conductive tissues and insects such as these, whose mouthparts cannot penetrate past it, are little affected by neem treatments. On the other hand, leafhoppers and planthoppers, that feed at least half the time on the deeper layer of conductive tissues (called the xylem), get knocked down.

5

Effects on Insects

The growing accumulation of experience demonstrates that neem products work by intervening at several stages of an insect's life. The ingredients from this tree approximate the shape and structure of hormones vital to the lives of insects (not to mention some other invertebrates and even some microbes). The bodies of these insects absorb the neem compounds as if they were the real hormones, but this only blocks their endocrine systems. The resulting deep-seated behavioral and physiological aberrations leave the insects so confused in brain and body that they cannot reproduce and their populations plummet.

Increasingly, approaches of this kind are seen as desirable methods of pest control: pests don't have to be killed instantly if their populations can be incapacitated in ways that are harmless to people and the planet as a whole. In the 1990s this is particularly important: many synthetic pesticides are being withdrawn, few replacements are being registered, and rising numbers of insects are developing resistance to the shrinking number of remaining chemical controls.

The precise effects of the various neem-tree extracts on a given insect species are often difficult to pinpoint. Neem's complexity of ingredients and its mixed modes of action vastly complicate clarification. Moreover, the studies to date are hard to compare because they have used differing test insects, dosages, and formulations. Further, the materials used in various tests have often been handled and stored differently, taken from differing parts of the tree, or produced under different environmental conditions.

But, for all the uncertainty over details, various neem extracts are known to act on various insects in the following ways:

- Disrupting or inhibiting the development of eggs, larvae, or pupae;
- Blocking the molting of larvae or nymphs;
- Disrupting mating and sexual communication;
- Repelling larvae and adults;
- Deterring females from laying eggs;

- Sterilizing adults;
- Poisoning larvae and adults;
- Deterring feeding;
- Blocking the ability to "swallow" (that is, reducing the motility of the gut);
- Sending metamorphosis awry at various stages; and
- Inhibiting the formation of chitin.[1]

As noted earlier, neem extracts have proved as potent as many commercially available synthetic pesticides. They are effective against dozens of species of insects at concentrations in the parts-per-million range. At present, it can be said that repellency is probably the weakest effect, except in some locust and grasshopper species. Antifeedant activity (although interesting and potentially extremely valuable) is probably of limited significance; its effects are short-lived, and highly variable. Blocking the larvae from molting is likely to be neem's most important quality. Eventually, this larvicidal activity will be used to kill off many pest species.

INSECTS AFFECTED

By 1990, researchers had shown that neem extracts could influence almost 200 insect species. These included many that are resistant to, or inherently difficult to control with, conventional pesticides: sweet potato whitefly, green peach aphid, western floral thrips, diamondback moth, and several leafminers, for instance.

In general, it can be said that neem products are medium- to broad-spectrum pesticides of plant-eating (phytophagous) insects. They affect members of most, if not all, orders of insects, including those discussed below.

Orthoptera

In Orthoptera (such as grasshoppers, crickets, locusts), the antifeedant effect seems especially important. A number of species refuse to feed on neem-treated plants for several days, sometimes several weeks. Recently, a new effect, which converts the desert locust from the gregarious swarming form into its nonmarauding solitary form, has been discovered.

---

[1] Chitin is the material comprising the insect's exoskeleton. Stopping the formation of a new "skin" for the next stage in its development is one way that azadirachtin acts to regulate the growth of an insect.

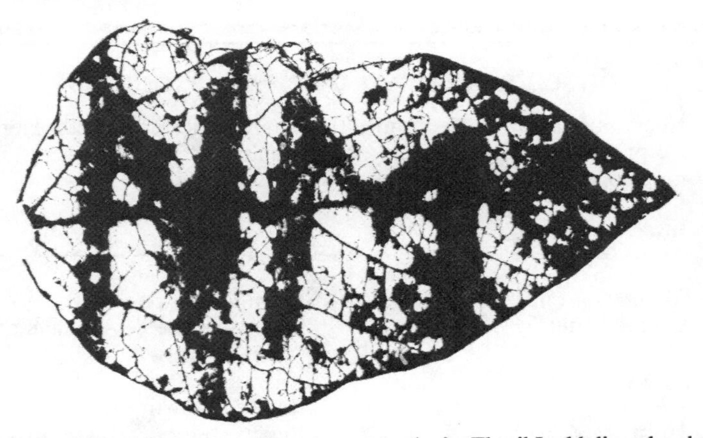

As a test of neem's ability to repel insects, entomologist Thyril Ladd dipped a glass rod into dilute neem extract and wrote the letters "N" and "M" on a soybean leaf. He then exposed the leaf to the Japanese beetle, a pest renowned for a voracious appetite for soybean leaves. As can be seen, the bulk of the leaf was stripped to its woody veins, but the insects succumbed to starvation rather than nibble on the "N" or "M." (T. Ladd)

Homoptera

Aphids, leafhoppers, psyllids, whiteflies, scale insects, and other homopterous pests are sensitive to neem products to varying degrees. For instance, nymphs of leafhoppers and planthoppers show considerable antifeedant and growth-regulating effects. However, scale insects (especially soft scale), are little affected. Phloem feeders, such as aphids, are in general not good candidates for neem used systemically (see earlier). In some cases, the host plant may influence the degree of control; this seems to apply to some whiteflies, which are affected on some crops but not on others.

Neem derivatives may also influence the ability of homopterous insects to carry and transmit certain viruses. It has been shown, for example, that low doses keep the green rice leafhopper from infecting rice fields with tungro virus. The cause is uncertain but seems to be only partly owing to neem killing the insects or modifying their feeding behavior.

Thysanoptera

Neem is very effective on thrips larvae, which occur in the soil. However, once the adult thrips and related pests have taken up residence on the plants themselves, they are less sensitive to neem extracts. Oily formulations have shown some success in exploratory trials (perhaps because the oil coated and suffocated these minute creatures).

Insects Affected by Neem Products

Neem is known to affect more than 200 species of insects. Here we present brief information on a sampling of them to show the range of effects and the range of species affected.

Insect	Effects
Mediterranean fruit fly	Disrupts growth, toxic
Oriental fruit fly	Arrests pupae development, retards growth, toxic to larvae
Face fly	Retards growth, toxic
Horn fly	Repels, retards growth, disrupts growth
Whitefly	Repels, retards growth, inhibits feeding
Housefly	Inhibits feeding, disrupts molting, repels
Sorghum shoot fly	Inhibits feeding
Yellow-fever mosquito	Kills larvae, disrupts molting
House mosquito	Toxic to larvae
Flea	Retards growth, repels, inhibits feeding, disrupts growth, eggs fail to hatch
Head lice	Kills, very sensitive to neem oil—traditional use in Asia
Spotted cucumber beetle	Retards growth, inhibits feeding
Mexican bean beetle	Retards growth, inhibits feeding, disrupts molting
Colorado potato beetle	Eggs fail to hatch, larvae fail to molt with azadirachtin levels as low as .3 ppm, inhibits feeding
Flea beetle	Inhibits feeding
Khapra beetle	Inhibits feeding, disrupts molting, toxic to larvae
Confused flour beetle	Inhibits feeding, disrupts molting, toxic to larvae
Japanese beetle	Repels, retards growth, inhibits feeding, disrupts growth
Red flour beetle	Inhibits feeding, toxic
American cockroach	Reduces fecundity and molts, reduces number of fertile eggs
Bean aphid	Reduces fecundity, disrupts molting
Rice gall midge	Toxic

Insect	Effects
Western thrips	Retards growth
Diamondback moth	Strongly suppresses larvae and pupae, retards growth, inhibits feeding
Webbing clothes moth	Inhibits feeding, disrupts molting
Gypsy moth	Retards growth, inhibits feeding, disrupts growth
Corn earworm	Retards growth, inhibits feeding, disrupts molting
Pink bollworm	Retards growth, inhibits feeding
Fall armyworm	Retards growth, repels adults, inhibits feeding, disrupts molting, toxic to larvae
Tobacco budworm	Inhibits feeding
Tobacco hornworm	Inhibits feeding, disrupts growth, toxic
Cabbage looper	Inhibits feeding
Leafminer	Retards growth, inhibits feeding, disrupts molting, toxic
Serpentine leafminer	High pupal mortality, retards growth, inhibits feeding, disrupts molting, toxic to larvae
Brown planthopper	Inhibits feeding, repellent, disrupts growth, mating failures and sterility
Green leafhopper	Inhibits feeding
Migratory locust	Stops feeding, converts gregarious nymphs into solitary forms, reduces fitness, adults cannot fly
House cricket	Disrupts molting
Large milkweed bug	Toxic, disrupts growth
Mealy bugs	Repels, inhibits feeding
Milkweed bug	Difficulty in escaping the "skin" of the last molt, disrupts molting
Fire ant	Inhibits feeding, disrupts growth
Boll weevil	Inhibits feeding
Cowpea weevil	Inhibits feeding, toxic
Rice weevil	Inhibits feeding, disrupts growth, toxic

Coleoptera

The larvae of all kinds of beetles—especially those of phytophagous coccinellids (Mexican bean beetle and cucumber beetle, for example) and chrysomelids (Colorado potato beetle and others)—are also sensitive to neem products. They refuse to feed on neem-treated plants, they grow slowly, and some (such as the soft-skinned larvae of the Colorado potato beetle) are killed on contact.

Lepidoptera

From numerous field trials (notably on various moths), it appears that larvae of most lepidopterous pests are highly sensitive to neem. Indeed, it seems likely that armyworms, fruit borers, corn borers, and related pests will become the main targets of neem products in the near future. Neem blocks them from feeding, although this effect is usually less important than the disruption of growth it causes.

Diptera

Many species of dipterous insects—fruit fly, face fly, botfly, horn fly, and housefly, for example—are targets for neem products. Mosquitoes, too, are a possibility.

Hymenoptera

The freely feeding and caterpillar-like larvae of sawflies are target insects as well. In this group, neem's antifeedant and growth regulatory effects are both important.

Heteroptera

The "true" bugs—including many pests such as the rice bug, the green vegetable bug, and the East African coffee bug that suck juices from crops and trees—are affected by neem products. Neem's systemic qualities affect their feeding behavior and disrupt their growth and development.

EXAMPLES

As discussed, neem's effects vary with different insects. Some effects on a small selection of major pests are summarized below.

Desert Locust

Recent laboratory research has shown that neem oil causes "solitarization" of gregarious locust nymphs.[2] After exposure to doses equal to a mere 2.5 liters per hectare, the juveniles fail to form the massive, moving, marauding plagues that are so destructive of crops and trees. Although alive, they became solitary, lethargic, almost motionless, and thus extremely susceptible to predators such as birds. Neem affects grasshopper nymphs similarly.

This discovery differs from earlier ones on locusts. Those first approaches used alcoholic extracts and were aimed at disrupting metamorphosis or at stopping adult locusts from feeding on crops. The new approach uses neem oil enriched with azadirachtin to prevent locusts from developing into their migratory swarms. It apparently blocks the formation of the hormones and the pheromones needed to maintain the yellow-and-black gregarious form, which plagues arid Africa and the Middle East. In an interesting aside, it has been shown that neem oil destroys their antennae, even when applied to the abdomen.

Neem trees grow well throughout the locust zones of Africa and the Middle East, and thus, in principle at least, the means to control the plagues could be locally produced.

Cockroach

Neem kills young cockroaches and inhibits the adults from laying eggs. Baits impregnated with a commercial preparation of neem-seed extract proved to retard the growth of oriental, brown-banded, and German cockroaches.[3] First-instar nymphs of all three species failed to develop, and all died within 10 weeks. Last-instar nymphs exhibited retarded growth, and half of them died within 9 weeks. After 24 weeks, only 2 out of the 10 surviving German-cockroach nymphs had reached adulthood.

In a "taste test," American cockroach adults preferred neem-treated pellets over untreated ones, but neem-treated milk cartons repelled them.[4]

Brown Planthopper

Neem cake (the residue left after oil has been removed from the kernel) has proved so successful that Philippine farmers are already

[2] Schmutterer and Freres, 1990.
[3] The baits were lab-chow pellets laced with Margosan-O® at 0.5 ml per pellet.
[4] Adler and Uebel, 1985.

using it on a trial basis against the brown planthopper (and other rice pests).[5] Neem oil is being employed as well. Five applications of a 25-percent neem-oil emulsion sprayed with an ultra-low-volume applicator is said to protect rice crops against this increasingly severe scourge. It has been estimated that one neem tree provides enough ingredients to protect a hectare of rice. This use alone exemplifies the economic importance of further developing the neem tree for pest control.[6]

Stored-Product Insects

Neem shows considerable potential for controlling pests of stored products. This is one of the oldest uses in Asia, and the literature contains many references to its benefits. In the traditional practice, neem leaves are mixed with grain kept in storage for 3–6 months. The ingredients responsible for keeping out the stored-grain pests are not yet identified—but they work well.

In this connection, repellency seems of primary importance. For instance, treating jute sacks with neem oil or neem extracts prevents pests—in particular, weevils (*Sitophilus* species) and flour beetles (*Tribolium* species)—from penetrating for several months. For this use, the degradation problem caused by sunlight is less of a concern because the products are mostly away from the sunlight, inside jars or other containers.

Neem oil is an extremely effective and cheap protection for stored beans, cowpeas, and other legumes. It keeps them free of bruchid-beetle infestations for at least 6 months, regardless of whether the beans were infested before treatment or not.[7] This process may be unsuited for use in large-scale food stores, but it is potentially valuable for household use and for protecting seeds being held for planting. The treatment in no way inhibits the capacity of the seeds to germinate.

Neem has also been used in India to protect stored roots as well as tubers against the potato moth. Small amounts of neem powder are said to extend the storage life of potatoes 3 months.

Armyworm

Azadirachtin has proved an effective prophylactic against armyworms at extremely low concentrations—a mere 10 mg per hectare.[8]

[5] Saxena et al., 1984; von der Heyde et al., 1984.
[6] Information from R.C. Saxena.
[7] The amount of oil used was 2–3 ml per kg of beans. Neem oil shows a strong ovicidal effect in bean-seed beetles (bruchids), but its sterilizing and other influences may also be important in controlling these pests, which constitute a major problem when storing beans of many types (Zehrer, 1984).
[8] Information from J. Klocke.

Left row: untreated white cabbage, badly damaged by diamondback moth (and aphids). Right row: cabbage treated with aqueous neem-seed-kernel extract is largely undamaged. On the diamondback moth, neem exerts a combination of effects: it repels, it deters oviposition (eggs that are laid never hatch), and it disrupts molting. This extremely serious pest is found worldwide and in some locations is playing havoc with vital crops of leafy vegetables such as cabbage. (H. Schmutterer)

For instance, it inhibits the fall armyworm, one of the most devastating pests of food crops in the western hemisphere. It has, however, been found necessary to treat the crop before the insects arrive. If this is done, they "march right on past the fields," but once they have taken up residence, it is harder to get them to move on.

Colorado Potato Beetle

In advanced trials in the United States, neem extracts have controlled the Colorado potato beetle.[9] This is a significant pest in North America and Europe that is becoming increasingly resistant to broad-spectrum insecticides.

In experiments in Virginia, for example, neem-seed extracts (at relatively low concentrations of 0.4 percent, 0.8 percent, and 1.2 percent) were tested in potato fields both with and without the synergist piperonyl butoxide (PBO). All treatments significantly lowered the potato beetle populations and raised potato yields; however, the extracts containing PBO were the most effective. The sprayings were most effective when the larvae were young, and were best when conducted as soon as the eggs hatched.[10]

Leafminers

When birch trees were sprayed to control the birch leafminer (*Fenusa pusilla*), neem extract seemed to perform as well as the registered commercial pesticide Diazinon®. It was, however, slower acting, and the insects continued to damage trees before they died. This leafminer is a serious pest in parts of North America, often browning the crowns of entire forests.

The U.S. Environmental Protection Agency has approved a neem-seed-extract formulation for use on leafminers. This commercial product, now available almost nationwide, is expected to be especially useful against those leafminers that attack horticultural crops. Added to the soil, neem compounds enter the roots and move up into the crop's leaves so that leafminers munching on the leaves get their molting-hormone jammed, and they end up fatally trapped inside their own juvenile skins.

European Corn Borer

The European corn borer, a highly adaptable pest of corn and other crops, was introduced to North America in 1917 and subsequently

---

[9] The statements here are based largely on research at Virginia Polytechnic Institute and State University, but generally similar results have been found in various parts of the United States and Canada.

[10] Lange and Feuerhake, 1984.

slashed Canada's corn yields in half. Today, it infests 40 million acres of corn in the United States each year, and in just an average year American farmers spend an estimated $400 million on chemicals to fight it.

Laboratory tests using neem products on this corn borer larvae produced 100 percent mortality at 10 ppm azadirachtin; 90 percent mortality at 1 ppm. Lower concentrations (0.1 ppm azadirachtin) left the larvae apparently unaffected, but the adults that later emerged had grossly altered sex ratios (there were many more males than females) and the few remaining females laid fewer eggs and laid them too late. This combination of effects suggests that azadirachtin could be effective for controlling this terrible pest.[11]

Mosquitoes

The larvae of a number of mosquito species (including *Aedes* and *Anopheles*) are sensitive to neem. They stop feeding and die within 24 hours after treatment. If neem derivatives are used alone, relatively high concentrations are required to obtain high mortality.[12] Nonetheless, the use of simple and cheap neem products seems promising for treating pools and ponds in the towns and villages of developing countries. In one test, crushed neem seeds thrown into pools proved nearly as effective at preventing mosquito breeding as methoprene, a rather expensive pesticide that is usually imported in developing countries.

Aphids

In the Dominican Republic, water extracts of neem seed proved effective against *Aphis gossypii* on cucumber and okra and against *Lipaphis erysimi* on cabbage.[13] This was in direct-contact sprays.

As noted earlier, neem extracts applied in a systemic manner (that is, within plants) usually have little effect on aphids. Apparently, this is because aphids feed only on the phloem tissues, where, for some unknown reason, neem materials accumulate least.

Fruit Flies

Fruit flies (including the notorious medfly) are among the most serious horticultural pests. They cause millions of dollars in damage to fruits, and their very presence in the tropics is keeping dozens of delicious fruits from becoming major items of international trade. But, at least in experiments, the medfly is proving susceptible to neem. This

---

[11] Arnason et al., 1985.

[12] This seems to be particularly true in the case of the yellow-fever mosquito, *Aedes aegypti*.

[13] Information from H. Schmutterer.

insect pupates underground, and in trials in Hawaii, spraying dilute neem solution under fruit trees resulted in 100 percent control.[14]

More important, the neem materials were compatible with the biological-control organisms (braconid wasps) used to control fruit flies. When neem was applied to soil at levels that completely inhibited the pest from emerging from pupation, the parasites developing in these pupae emerged and exhibited normal life spans and reproductive rates. Thus, neem is compatible with biological control of fruit flies. Diazinon®, the current soil treatment for fruit flies, kills not only fruit flies but their internal parasites as well.[15]

Gypsy Moth

The U.S. Environmental Protection Agency has approved a neem-seed-extract formulation for use on gypsy moth, a pest that is ravaging forests in parts of North America. In laboratory trials, a commercial neem formulation (Margosan-O®) produced 100 percent kill at very low concentrations (0.2 liters per hectare). After 25 days, the larvae were shrivelled, had stopped eating, and were dying. Field tests are in progress.

Horn Flies

Ground-up neem seed and stabilized neem extracts can prevent horn flies from breeding in cattle manure. In recent U.S. Department of Agriculture trials in Kerrville, Texas, cattle were fed a diet containing these neem materials in the feed. The animals readily consumed feed containing 0.1–1 percent ground neem seed. The neem compounds passed through the digestive tract and into the manure where they kept the fly larvae from developing.[16]

Blowflies

In Australia neem products have been tested against blowflies on sheep. The larvae of these pests penetrate and burrow under the skin of sheep. They are a major economic burden to Australia's farmers because many of the sheep die. In the tests, azadirachtin kept blowflies from "striking" (that is, laying their eggs on sheep).[17]

As a result of the excitement this discovery engendered, 1,000 hectares of neem have been planted in Queensland at a cost of more than $4 million. At least one Australian company has been established to produce and distribute neem products to sheep farmers.

---

[14] Information from J.D. Stark. The neem formulation (Margosan-O®) proved less effective than Diazinon® but at low levels (10 ppm azadirachtin in the soil) provided excellent control for the flies.
[15] Information from J.D. Stark.
[16] Information from J.A. Miller.
[17] Information from M.J. Rice.

6

Effects on Other Organisms

Although neem's effects on pestiferous insects are by far the best known, the tree's various products can influence other pest organisms as well. In the long run, these may well prove the most important of all. At present, however, the effects on noninsect pests are poorly understood. This chapter highlights some of the findings to date.

NEMATODES

Neem products affect various types of nematodes. This may be significant because certain of these thread worms are among the most devastating agricultural pests and are also among the most difficult to control. In addition, an increasing number of synthetic nematocides have had to be withdrawn from the market for toxicological reasons.

Today, there is a small but increasing body of evidence that neem might provide useful replacements. Certain limonoid fractions extracted from neem kernels are proving active against root-knot nematodes, the type most devastating to plants. They inhibit the larvae from emerging and the eggs from hatching, and in at least one test they have done so at concentrations in the parts-per-million range.[1] Water extracts of neem cake (the residue remaining after the oil has been pressed out of the seeds) are also nematocidal.

In a careful trial in Aligarh, India, amending soil with sawdust and neem cake dropped the root-knot index to zero and, of all the treatments tested, gave the greatest growth of tomatoes, a crop that is very sensitive to these nematodes.

In tests in a greenhouse and in the field in Germany, tomato plants were obviously improved by neem products, but there was no significant difference in the numbers of some nematode species in the soil. However, among treated and untreated soils the majority was extracted from the roots of plants in untreated soil.[2]

---

[1] Devakumar et al., 1985.
[2] Rössner and Zebitz, 1987a.

Cardamom growers in South India are already using neem cake to control nematodes. Of 19 growers interviewed recently, 17 said that nothing works as well. These were sophisticated farmers who monitor world cardamom prices regularly and use synthetic chemicals for controlling other pests in their fields. In other words, they weren't using neem out of ignorance or poverty. They incorporate 100–259 kg per hectare of neem cake in their cardamom fields every year. About 3,000 tons of neem cake are now used annually in India's Cardamom Hills. It is sold by pesticide dealers, who transport it from 250–300 km away.[3]

SNAILS

Various neem extracts kill snails. This appears to be beneficial in some cases.

In laboratory tests, for example, ethanol extracts proved toxic to the aquatic snail (*Biomphalaria glabrata*), a species that is necessary to the life cycle of the parasite causing schistosomiasis (bilharzia). The extracts killed both the adult snail and its eggs.[4] This raises the possibility that neem products may find a role in controlling schistosomiasis, a horrible scourge that infects some 200 million people in the tropics.

In another test, an aqueous solution of neem fruit resulted in a 100-percent kill of *Melania scabra*.[5] This snail, common throughout the Orient, is a vector of lung flukes, a parasitic flatworm that encysts in the lungs of livestock, wildlife, and people, causing debilitation and sometimes death.

CRUSTACEANS

Little is known about neem's effects—beneficial or detrimental—on crustaceans. However, in one intriguing set of experiments in the Philippines, it proved beneficial.

In rice paddies, the ostracod *Heterocypris luzonensis* feeds on the blue-green algae that fix nitrogen from the air. This minute aquatic crustacean thereby reduces a source of fertilizer for the crop. Killing this tiny creature thus would indirectly boost the nitrogen available and probably increase rice yields.[6] Aqueous neem-kernel extracts have killed it very effectively under laboratory conditions.[7]

---

[3] Information from S. Ahmed.
[4] Information from D. Heyneman.
[5] Muley, 1978.
[6] Ketkar and Ketkar, 1984.
[7] Grant and Schmutterer, 1986.

FUNGI

Neem has demonstrated antifungal activity. Should this prove widely applicable, the availability of a natural fungicide that can be grown, extracted, and applied by farmers themselves could be of great consequence to worldwide agriculture and food supply. Fungi attack crops in countless numbers and forms. They are constantly evolving enemies of farms and forests. Many can reach epidemic proportions, a few have no cures, and some can make certain crops impossible to grow. And, despite the best of modern science, they still threaten wheat, corn, rice, and other plants that feed the world.

Not a lot is known about neem's practical use against rots, smuts, wilts, mildews, die-backs, and other fungal plant diseases. However, several tests have indicated considerable promise.

In one test, neem oil protected the seeds of chickpeas against the serious fungal diseases *Rhizoctonia solani*, *Sclerotium rolfsii*, and *Sclerotinia sclerotiorum*. It also slowed the growth of *Fusarium oxysporum* but did not kill it. In addition, neem cake incorporated into the soil completely blocked the development of the resting forms of *R. solani*—thereby interfering with the long-term survival of this devastating fungus.[8]

In another, neem-seed extracts showed beneficial effects against leaf fungi. Spraying crude neem oil on lilac bushes, when done before any sign of outbreak, prevented powdery mildew from breaking out for the rest of the season. This protectant also gave essentially 100 percent control on hydrangeas in greenhouses—better than Benlate® (benomyl), the standard mildew treatment in much of the world.[9]

In the case of bean rust, neem extracts have given 90 percent control when applied before the plants were exposed to the fungus. However, they worked poorly once rust was established.[10]

In addition to affecting root-knot nematodes, treating soil with neem can reduce the populations of pest fungi in the rhizosphere that attack and feed off plant roots.[11]

Aflatoxin

A truly unusual and potentially notable connection between neem and fungi has recently been reported from Louisiana.[12] In trials there, neem-leaf extract failed to kill the fungus *Aspergillus flavus*, but, against all expectations, it completely stopped it from producing aflatoxin (see sidebar).

---

[8] Singh et al., 1980.
[9] Information from J. Locke.
[10] Locke, 1990.
[11] Singh et al., 1985.
[12] Information in this section is from D. Bhatnagar. For more, see sidebar.

The Cancer-Causing Fungus

While growing up in Rajasthan, India, Deepak Bhatnagar was impressed with neem's qualities. He often saw his parents using the leaves to keep insects out of the wheat they stored in their home. He also saw how well these leaves worked against skin infections when they cured a persistent ulcer on his leg—one that had baffled the best of medical practitioners.

Today, Bhatnagar works at the U.S. Department of Agriculture and has taken up the study of neem's effects on certain fungi. In tests in his laboratory in New Orleans, he ground up (or boiled) neem leaves in water (or in potassium-phosphate-buffered solution to remove any possible pH effect) and applied the resulting solutions to *Aspergillus flavus*. This fungus, one of the most deadly on earth, grows on various foods and produces chemicals called aflatoxins that are highly carcinogenic. When Bhatnagar looked at the fungal cultures four days later, they seemed normal. But when he tested them chemically, he could find only 2 percent of the aflatoxin that the fungus normally would have produced. Neem had left the microbe alive, but had switched off its ability to produce aflatoxins.

Bhatnagar then moved on to greenhouse studies. He injected neem solutions into cotton bolls, and later infected the bolls with the fungus. Again, aflatoxin production was inhibited (see diagram opposite).

Experiments are now under way to determine which components in the neem leaves are responsible for the bioactivity. Once they are identified, cost-effective and efficient delivery systems can probably be developed to control aflatoxin synthesis by the fungus on various crops.

Bhatnagar says that the results are "promising but preliminary." But if his work proves that neem is safe and effective for aflatoxin control, it may open the door to a simple, inexpensive method for protecting stored foods using locally produced materials, even in the remotest rural villages.

This is especially significant these days. With the availability of ever more sensitive chemical analyses, health officials are becoming alarmed at aflatoxin's widespread occurrence and potential hazard. Anything that might protect food supplies in tropical regions that can ill afford synthetic fungicides and have difficulty keeping foods fungus-free could be of immense significance.

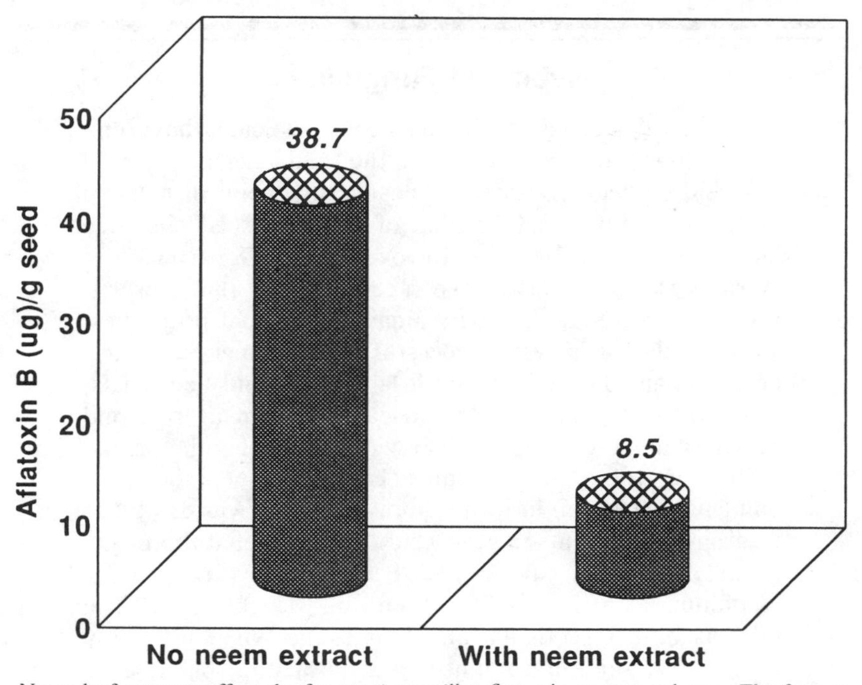

Neem-leaf extracts affect the fungus *Aspergillus flavus* in an unusual way. The fungus appears to grow normally, but it produces far less aflatoxin. This is important because aflatoxin is a powerful carcinogen that is of increasing concern in the world's food supplies. The levels shown here were measured in samples of the fungus grown on developing cottonseed. (D. Bhatnagar)

Greenhouse studies have since confirmed these laboratory findings. The extracts appear to halt the formation of substances called polyketides, which the fungi convert into aflatoxin. The enzymes for the conversion remain in place, but key chemicals they need to synthesize the feared toxin are no longer available.

It proved easy to take advantage of this in practice: neem leaves were mashed in water, the liquid separated, and it was applied without further refinement. This crude liquid extract turned off aflatoxin production in both laboratory cultures and cotton bolls on living plants.

These findings could be of immense significance. Aflatoxin causes liver cancer, and under hot and humid conditions, where fungi thrive, it can form on peanuts, corn, cottonseed, and other widely eaten food crops. It is of great concern these days; it not only threatens health, it also promises economic catastrophe. For example, the United States may soon be banned from exporting cottonseed to feed Europe's cattle. The U.S. aflatoxin limit is 20 parts per billion (ppb), but Europe's goal for the future is a mere 2 ppb in feed and only 0.5 ppb in milk. Also, aflatoxin-contaminated local foods and feeds are causing increased concern in Asia and Africa.

Neem Oil Fungicides

So far, almost all studies on neem pesticides have employed limonoids extracted from the seed kernel. Neem oil has seldom been considered because its chemical makeup is not very different from that of common seed oils such as soybean or olive oil. However, U.S. Department of Agriculture researchers have recently found that, surprisingly, neem oil has its own valuable pesticidal properties. In particular, it is very successful against fungi that cause certain plant diseases. In both laboratory and field trials, neem oil has controlled the diseases known as rust and powdery mildew—and it did so without harming the plants.

In the first of these trials, James Locke, a plant pathologist, emulsified neem oil in water, sprayed it over various types of ornamental plants in pots, and then subjected the plants to rust or powdery mildew. "We had success with emulsions containing as little as 0.25 percent oil," Locke says. "The oil was both insecticidal and fungicidal. We don't really know why, since it contains no azadirachtin, but it does work."

In greenhouse trials, plant pathologist J. Rennie Stavely found that neem oil is nearly 100 percent effective against rust on beans. Although its effectiveness was slightly less dramatic on bean plants in the field, neem oil still reduced this serious fungal disease enough to be cost-effective.

Hydrangea leaves exposed to powdery mildew. One (left) became badly infected with the fungus. The other, protected by a dilute solution of need-seed oil in water, grew to full size and was almost unaffected. (J. Locke, Agricultural Research Service, USDA)

PLANT VIRUSES

Plant viruses pose some of the most severe threats to world agriculture. Because they invade the crop's cells and cloak themselves with the plant's normal life processes, they are far more difficult to control than free-living organisms such as bacteria, protozoa, or fungi. At present, we can only try to halt their spread—something nearly impossible to achieve under even the best of circumstances—because viruses "hitch rides" in insects such as aphids, as well as on dirty tools, blowing dust, or spreading floodwaters.

A few virus-inhibiting chemicals are known for treating human and animal diseases (AZT for AIDS, for instance), but at present, none are available for treating plants. Neem might be the first. Crude extracts seem to bind certain plant viruses effectively, and so limit infection.[13]

However, for the moment at least, neem seems most effective at interfering with the transmission of plant viruses carried by insects. This conclusion is drawn from several successful tests of neem's effects against insect vectors of plant viruses.

These tests include the following:

- A trial in the Philippines where rice fields sprayed with neem oil had significantly lower incidence of the ragged-stunt virus, which affects rice and is transmitted by the brown planthopper;[14]
- A second trial in the Philippines where mixtures of neem oil and custard-apple oil interfered with the transmission of tungro virus, another rice pest;[15]
- Experiments in India where neem-leaf extracts reduced the transmission of tobacco mosaic, a virus that seriously affects several vegetable crops;[16]
- Field trials in the Philippines where fields treated with urea and neem cake were found to be lower in viral diseases than those treated with urea alone;[17] and
- Enzyme-linked immunosorbent assays showing that rice seedlings grown in soil treated with neem cake were significantly freer of rice tungro viruses (transmitted by green rice leafhopper) than those in untreated control plots.[18]

On the other hand, not all trials have been this successful. In the United States, daily applications of neem leaf extracts over a month's time to turnip plants infected with cauliflower mosaic virus did *not* reduce viral infection.[19]

---

[13] Singh, 1971; Tripathi and Tripathi, 1982.
[14] Saxena et al., 1981.
[15] Mariappan and Saxena, 1984. For information on custard apples, see the companion report *Lost Crops of the Incas*.
[16] Information from B.G. Joshi.
[17] Saxena et al., 1984.
[18] Saxena et al., 1987.
[19] Information from D.W. Unander.

NONTARGET SPECIES

As has been mentioned, neem extracts proved to be "soft" on unintended targets. Further examples follow.

Earthworms

In greenhouse studies, when neem leaves and seed kernels were incorporated into potting soil containing earthworms (*Eisenia foetida*), the number of young worms produced increased 25 percent.[20] In field trials there were no differences in the number of worms, but the average weight of each worm was highest in neem-treated plots. Thus, it seems possible that neem products can favor earthworms, at least under certain conditions.

Beneficial Insects

Neem seems remarkably benign to spiders, butterflies, and insects such as bees that pollinate crops and trees, ladybugs that consume aphids, and wasps that act as parasites on various crop pests.[21] In the main, this is because neem products must be ingested to be effective. Thus, insects that feed on plant tissues succumb, while those that feed on nectar or other insects rarely contact significant concentrations of neem products.

All this is coming clearer from recent research. For example, only after repeated spraying of highly concentrated neem products onto plants in flower were worker bees at all affected. Under these extreme conditions, the workers carried contaminated pollen or nectar to the hives and fed it to the brood. Small hives then showed insect-growth-regulating effects; however, medium-sized and large bee populations were unaffected.[22]

Under laboratory conditions the larvae of ladybugs and lacewings have shown some insect-growth-regulating effects from neem picked up from the bodies of other insects. However, in greenhouse trials in Florida, neem products proved essentially nontoxic to predators and parasitoids of the cotton aphid and the sweet potato whitefly. Neither the amount of predation nor of parasitism was notably reduced.[23]

---

[20] Rössner and Zebitz, 1987b.
[21] Saxena, 1987.
[22] Schmutterer and Holst, 1987.
[23] Hoelmer et al., 1990.

A census of natural aphid enemies collected from seven different field trials indicated that neem has no detrimental effects on either predators (coccinellids, chrysopids, syrphids) or parasitoids (ichneumonids, braconids). The aphids in the neem-treated plots were actually carrying more parasites than were those in either the control plots or the plots treated with the insecticide pyrethrum.[24]

---

[24] Isman et al., in press.

7
Medicinals

Since antiquity neem has been renowned for healing. The earliest Sanskrit medical writings refer to the benefits of its fruits, seeds, oil, leaves, roots, and bark.[1] Each of these has long been used in the Indian Ayurveda and Unani systems of medicine. Thus, over thousands of years, millions of Asians have used neem medicinally. In addition, in places where the tree has been introduced in recent times, such as tropical America and Africa, it has also established a reputation as a useful cure for various ailments.

Today, the best-established and most widely recognized uses are based on its merits as a general antiseptic. Neem preparations are reportedly efficacious against a variety of skin diseases, septic sores, and infected burns. The leaves, applied in the form of poultices or decoctions, are also recommended for boils, ulcers, and eczema. The oil is used for skin diseases such as scrofula, indolent ulcers, and ringworm.

Cures for many more ailments have been claimed but have not been independently confirmed by trials under controlled conditions. Nonetheless, there are intriguing indications that neem might in future be used much more widely. These promising, but unproved, applications include anti-inflammatory, hypotensive, and anti-ulcer treatments.

A summary of some recent results in medical and veterinary studies follows.

FUNGICIDES

Neem has proved effective against certain fungi that infect the human body. Such fungi are an increasing problem and have been difficult to control by synthetic fungicides. For example, in one laboratory study,[2]

---

[1] The tree's Sanskrit name was "arishtha," meaning "reliever of sickness."
[2] Khan and Wassilew, 1987.

neem preparations showed toxicity to cultures of 14 common fungi, including members of the following genera:

- *Trichophyton*—an "athlete's foot" fungus that infects hair, skin, and nails;
- *Epidermophyton*—a "ringworm" that invades both skin and nails of the feet;
- *Microsporum*—a ringworm that invades hair, skin, and (rarely) nails;
- *Trichosporon*—a fungus of the intestinal tract;
- *Geotrichum*—a yeastlike fungus that causes infections of the bronchi, lungs, and mucous membranes; and
- *Candida*—a yeastlike fungus that is part of the normal mucous flora but can get out of control, leading to lesions in mouth (thrush), vagina, skin, hands, and lungs.

ANTIBACTERIALS

In trials neem oil has suppressed several species of pathogenic bacteria, including:

- *Staphylococcus aureus*.[3] A common source of food poisoning and many pus-forming disorders (for example, boils and abscesses), this bacterium also causes secondary infections in peritonitis, cystitis, and meningitis. Many strains are now resistant to penicillin and other antibiotics, one reason for the widespread occurrence of staphylococcal infections in hospitals.
- *Salmonella typhosa*.[4] This much-feared bacterium, which lives in food and water, causes typhoid, food poisoning, and a variety of infections that include blood poisoning and intestinal inflammation. Current antibiotics are of only uncertain help in treating it.

However, neem has many limitations as an antibiotic. In the latter test, neem showed no antibacterial activity against certain strains of the above bacteria, and none against *Citrobacter*, *Escherichia coli*, *Enterobacter*, *Klebsiella pneumoniae*, *Proteus mirabilis*, *Proteus morgasi*, *Pseudomonas aeruginosa*, *Pseudomonas* E01, and *Streptococcus faecalis*.

ANTIVIRAL AGENTS

In India, there is much interesting, but anecdotal, information attributing antiviral activity to neem. Its efficacy—particularly against

[3] Schneider, 1986.
[4] Patel and Trivedi, 1962.

pox viruses—is strongly believed, even among those of advanced medical training. Smallpox, chicken pox, and warts have traditionally been treated with a paste of neem leaves—usually rubbed directly onto the infected skin.

Experiments with smallpox, chicken pox, and fowl pox suggest that there may be a true biological basis for this practice. Crude neem extracts absorbed the viruses, effectively preventing them from entering uninfected cells.[5] Unfortunately, no antiviral effects were seen once the infection was established within the cell. Thus neem was effective prevention, but not cure.

Recent pharmacological studies have supported the belief that neem leaves possess some antiviral activity. So far these are only preliminary and unconfirmed results, but they are intriguing, nonetheless. In the United States, aqueous neem-leaf extracts have shown low to moderate inhibition of the viral DNA polymerase of hepatitis B virus.[6] In Germany, an ethanolic neem-kernel extract has proved effective against herpes virus.[7] And in horticultural studies, crude extracts also seemed to effectively bind certain plant viruses, and so limit infection (see Chapter 6).

Should these early results prove to be soundly based, an array of extremely virulent and difficult diseases of people—not to mention of wildlife and livestock—might be treated.

DERMATOLOGICAL INSECTS

Given all of neem's insecticidal properties, it is perhaps not unexpected that it is a common folk remedy against maggots and head lice. In Haiti, for instance, crushed leaves are rubbed into open wounds that have become maggot infested. And in India and Bangladesh, villagers apply neem oil to the hair to kill head lice, reportedly with great success.

DENTAL TREATMENTS

As noted earlier, both in India and Africa millions of people use twigs as "toothbrushes" every day. For many the twig is neem. Dentists have endorsed this ancient practice, finding it effective in preventing periodontal disease.[8] It is unclear whether the benefit is due to regular gum massage, to preventing plaque buildup, to neem's inherent antiseptic action, or to all three.

---

[5] Rao et al., 1969; Rae and Sethi, 1972.
[6] Information from D.W. Unander.
[7] Information from H. Schmutterer.
[8] M. Elvin-Lewis, 1980; Henkes, 1986.

People using neem sticks as disposable toothbrushes are commonly seen in rural regions of South Asia and Africa. Research has found that neem twigs contain antiseptic ingredients and they are able to maintain healthy teeth and gums. (M. Elvin-Lewis)

As also noted earlier, a German company uses neem (actually, extracts of bark) as the active ingredient in toothpastes and other oral hygiene preparations. It claims that its tests prove neem bark to be highly effective at both preventing and healing gum inflammations and periodontal disease.

CHAGAS' DISEASE

Extracts of neem reportedly affect the kissing bugs that transmit the much-feared Chagas' disease (see sidebar, page 64). They do not kill the insect; instead they "immunize" it against parasites that live inside it for part of their life cycle.[9] The discovery may point the way to

[9] Information from H. Rembold and E.S. Garcia.

Chagas' Disease

About 20 million Latin Americans are infected with Chagas' disease; many live as helpless cripples, unable to work or enjoy life. A parasite (*Trypanosoma cruzi*) causes this major health problem. It lives and reproduces inside nerve and muscle cells, particularly those of the heart, and drains its victims of all their energy.

The parasite—a trypanosome related to the well-known one that causes the dreaded sleeping sickness in Africa—is spread by the so-called "kissing bugs." Similar to large bedbugs, these insects inhabit cracks and crevices in the walls and roofs of huts and houses in rural areas. Emerging at night, they bite and suck the blood of any sleeping people, pets, or livestock they encounter.

In this way kissing bugs pick up the parasite, but they pass it on not through their bite, but through their droppings. The parasite develops and multiplies within the bug's hindgut, and, in its infective stage, passes out with the excrement. Kissing bugs often defecate while they are feeding, and when the victim wakes up and scratches the itchy bump, the excrement, together with the parasite's infective stage, is rubbed into the wound and enters the bloodstream.

So far, there is no truly satisfactory control for this dread disease, but laboratories in Germany and Brazil have recently produced the makings of a possible breakthrough involving neem.

At the Max Planck Institute for Biochemistry in Martinsried, Germany, Heinz Rembold has been raising one species of kissing bug, *Rhodnius prolixus*, to study the effects of azadirachtin on its hormone system. He found that azadirachtin prevents young kissing bugs from molting and therefore from ever maturing, and it interferes with the adults' ability to reproduce.

The other prong of this research is being done at the Oswaldo Cruz Institute in Rio de Janeiro, Brazil. There, Eloi Garcia and his research team keep both parasite-free and parasite-infected kissing bugs. Recently, when they fed blood laced with azadirachtin to a group of infected bugs, they made a surprising discovery: 20 days later, the bugs were parasite free.

Rembold and Garcia believe that azadirachtin somehow disrupts the carefully synchronized arrangement that allows the parasite to develop and multiply within the insect's gut. They found that even a few micrograms of azadirachtin taken up with a blood meal proved effective and that the bugs remained free of infection even 20 days after they had eaten blood laden with parasites.*

* Garcia et al., 1984.

controlling this major health problem in Latin America, although delivering neem materials to these tiny bloodsuckers in rural hovels is perhaps impossible in practice.

The research was done both in Germany and Brazil, and has shown that feeding neem to the bugs not only frees them of parasites, but azadirachtin prevents the young insects from molting and the adults from reproducing.

The researchers conclude that azadirachtin does not simply kill the parasite because they can dose it with azadirachtin and it remains infectious. In any case, they point out that the bug excretes most of the azadirachtin within a few minutes of eating it.

Even with this short exposure, however, azadirachtin somehow disrupts the delicate host-parasite relationship. According to the research leader: "The bug has been affected in such a fundamental way that it is no longer attractive to, or a viable host for, the parasite."

Trypanosomes like these have always been extremely difficult to control. Like the AIDS virus, they constantly "change their coat," so that creating vaccines against them is difficult at best. Whether neem will provide the key to their control is uncertain, but, at the very least, it may be a valuable research tool for understanding the basis of the relationship between the host and the powerful parasite it passes on to millions of people.

MALARIA

Practitioners of the Indian Ayurveda medicine system have been preparing neem in oral doses for malarial patients for centuries.[10] Neem's antimalarial activity was reported in Ayurveda books as far back as 2000 B.C. (by Charaka) and 1500 B.C. (by Sushruta). Even outside India—in Nigeria and Haiti, for example—neem-leaf teas are used to treat malaria.

[10] This is normally done by decocting or extracting about 500 g of leaves in alcohol or water.

People as Pests

Sticking forlornly out of the ground, only five leaves left on its skimpy branches, the seedling is dwarfed by the placard proclaiming its name and planting date: *Azadirachta indica*, 15 November 1986. A neem tree.

According to H.J. von Maydell's field guide *Trees and Shrubs of the Sahel*, the neem is "fast-growing: two-thirds of the height (to 20 meters) may be reached after three years." Photos from India show an evergreen with well-leafed branches forming a thick canopy. Yet this specimen in dryland Africa is barely one meter tall, a twig holding grimly onto life in what should be the favorable environment of an agroforestry research station.

"What happened to that little thing?" asks a visitor. The reply is revealing: "It got browsed by too many people. It was planted as part of a trial, but after a few months the farm manager noticed it wasn't growing. Other trees around it were doing well, so he knew it wasn't soil deficiency or lack of rain. He thought it must be animals eating the leaves, so he fenced it in with wire mesh. There are dik-diks (small antelopes) and rabbits here, and he concluded they really liked neem leaves."

"But the tree still didn't grow. Finally he decided to watch, and found farm workers were picking the leaves—for medicine. In Kiswahili, the tree is called mwarubaini, which means '40 cures.' It's said to be able to cure 40 diseases, and everybody wanted some of that medicine. He had to order them not to touch the tree.

"And that's the only reason it's still alive."

Tom Pawlick
Agroforestry Today 1(2):2–5

In the past, researchers were unable to confirm that neem products can affect the malaria parasite *Plasmodium falciparum*. And it was not for want of trying. Various groups researching antimalarials repeatedly tested neem. The results—in infected mice, ducks, and chickens—were inconsistent and usually negative.

Nonetheless, there is recent evidence that improper extraction methods may explain the earlier failures. Certain extracts of neem leaf and neem seed have now proved effective against the malarial parasite, and the structure of one active component has been determined.[11] This compound, gedunin, is another limonoid. It is said to be as effective as quinine on malaria-infected cell cultures.

In India, it was recently reported that components of the ethanol extract of neem leaves and seeds were effective against chloroquine-sensitive and chloroquine-resistant strains of the malaria parasite.[12] Although all the different extracts tested suppressed the growth of parasites within 72 hours, the most potent were the ethanol extracts of neem leaves and the medium-polar extracts of neem seeds.

Although these are preliminary results, they indicate a potentially valuable line of research. Malaria is creeping back into areas where it had been eliminated earlier this century. It now infects approximately 110 million people annually, causing up to 2 million deaths. Moreover, there is a growing problem of resistance to the conventional treatments.

On the other hand, caution is also needed. As noted earlier, anecdotal evidence from Nigeria suggests that drinking neem-leaf teas over an extended period may lead to liver damage.

PAIN RELIEF AND FEVER REDUCTION

Neem may also be a ready source of low-cost analgesic (pain-relieving), or antipyretic (fever-reducing) compounds. It is used for these purposes everywhere it is grown. In trials, positive results have been obtained for significant analgesic, antipyretic, and anti-inflammatory effects. This may explain its wide use for treating fevers in general. Some of the anti-inflammatory compounds have even been patented.[13]

BIRTH CONTROL

Research has shown that neem oil acts as a powerful spermicide. This finding is preliminary and may eventually prove of little consequence, but it may also prove of paramount importance. Perhaps 80 percent of the expected population explosion, which may double the number of people on earth in the next 40 years, will occur in countries where neem can be grown. An inexpensive birth-control method that can be produced in the backyards of even the remotest and poorest villages could be a vital resource.

---

[11] Khalid et al., 1986; Khalid et al., 1989.

[12] Badam et al., 1987.

[13] Terumo Corporation 1983, 1985. Anti-inflammatory polysaccharides from *Melia azadirachta*. *Japan Kokai Tokyo Koho* JP 82-05532 and 83-225021.

Neem and Birth Control

Scientists at India's Defence Institute of Physiology and Allied Sciences (DIPAS) have isolated a neem-oil extract (Nim 76) that they believe can be refined into a new birth-control method. Their trials have found that neem oil is strongly spermicidal. Rhesus monkey and human spermatozoa, for example, became totally immotile within 30 seconds of contacting the oil.

Studies in 20 rats, 8 rabbits, 14 rhesus monkeys, and 10 human volunteers showed that neem oil applied intravaginally before sexual intercourse prevented pregnancy. Histopathological studies on the rat tissues (vagina, cervix, and uterus) showed no ill effects. By contrast, nonyl-phenoxy polyethoxy ethanol, the spermicide in a popular vaginal contraceptive cream, produced obvious irritation. Radioisotope studies indicated that neem oil was not absorbed from the vagina.

DIPAS scientists maintain that neem oil might be an ideal contraceptive: it is a natural product, readily available, inexpensive, and nontoxic. Moreover, they anticipate that it will be widely accepted. The only disadvantage, they say, is neem oil's unpleasant odor. However, adding a small amount of scent masks most of the smell without reducing the spermicidal property.

All in all, they conclude, neem oil has particular potential for widespread use by the poor. It seems likely to be the cheapest contraceptive available, and villagers in remote areas may well accept it as a regular method of birth control because sophisticated methods are financially beyond their reach and because they are extremely apprehensive about sterilization and other sophisticated methods.

In a scientific article in the *Indian Journal of Medical Research*, DIPAS researchers report that tests also show that Nim 76 can prevent a fertilized egg from implanting in the wall of the uterus. Nim 76 was effective in rats and rabbits if applied on day 2 to day 7 of the expected pregnancy. The minimum effective dose was very small (only 25 μl for rats). And a month after the applications ceased, the animals were completely fertile again. According to the researchers, there were no deleterious effects on subsequent pregnancies or offspring (see Appendix B, page 104).

Indian scientists have demonstrated that neem oil is a potential new contraceptive for women (see sidebar). It kills spermatozoa within 30 seconds and has proved effective both in laboratory trials and in practice—where an intravaginal dose of 1 ml of neem oil was used. Histopathology failed to reveal any side effects.

As a follow-up to these experiments, the Indian army provided neem oil to 20 soldiers and their families as a birth-control measure. This trial was considered so successful that the colonel in charge of the program was honored by the prime minister. A neem-oil formulation called "Sensal" is now sold in India for contraceptive purposes.

Neem-leaf extracts have also shown promise as male birth-control products because they reduce fertility in a variety of male mammals. Reportedly, there was no impotence or loss of libido.[14]

More details of these tentative but potentially far-reaching discoveries are given both in the sidebar (opposite) and in Appendix B.

VETERINARY MEDICINE

Ancient practice and initial testing of neem derivatives against various livestock pests indicated that this is an area of particular promise for the future. Some possibilities are discussed below.

Controlling Insects

Insects of veterinary importance are obvious targets for neem products. Some examples follow:

- **Maggots** Indians have traditionally crushed neem leaves and rubbed them into open wounds on cattle to eliminate maggots.
- **Horn flies** As noted in chapter 5, azadirachtin passes through the ruminant digestive tract and remains long enough that horn flies will not develop in the manure.
- **Blowflies** As also noted earlier, neem oil and neem-seed extract deterred the female blowfly, *Lucilia sericata*, from laying its eggs on sheep. Moreover, in Sri Lanka the oil is rubbed on cattle as a fly repellent.[15]
- **Biting flies** Azadirachtin also exerts an ovicidal effect in eggs of the blood-sucking fly *Stomoxys calcitrans*.[16]

---

[14] Sadre et al., 1984.
[15] Ganesalingham, 1987.
[16] Gill, 1972. In one trial, neem-kernel dust incorporated into the diet of the larvae resulted in 100 percent mortality within 2 days at a 10 percent treatment level. At 5 percent treatment, 91 percent of larvae died within 2 days, and 100 percent mortality was reached in 7 days.

Controlling Bacteria

The *Staphylococcus aureus* bacterium, mentioned earlier, also causes mastitis (inflammation of the mammary glands) in cows. Neem's apparent ability to control certain strains of this bacterium may thus be of great economic importance to dairying in the nations where neem grows.

Also, the salmonella bacterium, in addition to affecting people, causes abortion in horses, cattle, and sheep, as well as a variety of infections in poultry and livestock.

Controlling Intestinal Worms

Trials in Germany showed that neem also works against intestinal nematodes in animals.[17]

A WORD OF CAUTION

Medicines from plants should, of course, be treated with the same caution as medicines from laboratories. Neem oil seems to be of particular concern. Consuming it, although widely practiced in parts of Asia, is *not* recommended. Doses as small as 5 ml have killed infants,[18] and animal studies showed acute toxicity at doses as low as 14–24 ml per kg of body weight.[19] It seems possible that this was caused by contaminants rather than by the oil itself. In Germany, toxicological tests using oil obtained from clean neem kernels resulted in no toxicity, even at a concentration of 5,000 mg per kg of body weight in rats. Nonetheless, caution is called for.

The leaves or leaf extracts also should not be consumed by people or fed to animals over a long period. There are anecdotal reports of renal failure in Ghanaians who were drinking leaf teas as a malaria treatment.

None of this should be confused with earlier statements. The compounds and seed-kernel extracts responsible for the insecticidal activity appear to be essentially nontoxic to mammals (see Appendix A).[20]

---

[17] Information from H. Schmutterer. To be effective, the preparations had to have relatively high concentrations of azadirachtin.
[18] Sinniah and Baskaran, 1981.
[19] Gandhi et al., 1988.
[20] Koul et al., 1990.

8

Industrial Products

Beyond all the possible pesticides and pharmaceuticals, neem provides many useful and valuable commonplace materials. For instance, oil extracted from the seeds goes into soaps, waxes, and lubricants, as well as into fuels for lighting and heating. The solid residue left after the oil is removed from the kernels is employed as a fertilizer and soil amendment. In addition, wood from the trees is valued for construction, cabinetry, and fuel. The bark is tapped for gum and extracted for tannins and dental-care products. The leaves are sometimes used for emergency livestock feed. And the profuse flowers are a prized source of honey.

NEEM OIL

Of all these products, the oil is perhaps the most commercially important. In composition, it is much like other vegetable oils, composed primarily of triglycerides of oleic, stearic, linoleic, and palmitic acids.[1]

To obtain neem oil, the seeds are first broken open and the kernels separated. The kernels are then pressed in industrial expellers or in hand- or bullock-operated wooden presses (*ghanis*). The oil yield is sometimes as high as 50 percent of the weight of the kernel.

This "cold-pressed oil" is mainly used in lamps, soaps, and other nonedible products. It is generally dark, bitter, and smelly. Unlike most vegetable oils, it contains sulfur compounds, whose pungent odor is reminiscent of garlic.

A large industry in India extracts the oil remaining in the seed cake using hexane. This solvent-extracted oil is not as high quality as the

[1] The actual fatty-acid contents measured recently were 41 percent oleic, 20 percent stearic, 20 percent linolenic, 18 percent palmitic, and 1 percent linolenic. (Information from Neem Project, Justus-Liebig University, Germany.)

In India, neem ingredients are found in many popular consumer goods. Neem oil, for instance, has been a major ingredient in soaps for at least 50 years. Its antiseptic properties have been used to particular advantage in the manufacture of special medicated soaps and toothpastes. In addition, pharmaceutical preparations—emulsions, ointments, poultices, and liniments, as well as cosmetics such as creams, lotions, shampoos, hair tonics, and gargles—have been prepared. The latest cosmetic preparations are entirely free of the odor that previously restricted neem oil's use. (Agricultural Research Service, USDA)

cold-pressed oil, but it also goes into certain soaps and consumer products.

Purifying neem oil is an elaborate and costly process at present. In one method, the smelly sulfur compounds are distilled off, which frees the oil from both odor and susceptibility to rancidity (because it also removes the free fatty acids). This process has long been used industrially.

As an alternative to pressing out the oil, the kernels can be extracted first with alcohol and then with hexane.[2] Alcohol removes the bitter and odoriferous compounds; hexane recovers the oil. This stepwise extraction upgrades both meal and oil. On the other hand, it requires

[2] This combination process has been developed at the Regional Research Laboratory in Hyderabad, India.

Neem Oil in India

The neem trees occurring throughout India represent a large, although very scattered, resource. Already, neem oil is a common commodity traded freely in the markets, but much more could be produced. It has been estimated that India's neems bear about 3.5 million tons of kernels each year and that, in principle, about 700,000 tons of oil might be recoverable. The annual production in the late 1980s was only around 150,000 tons. (About 34 tons of neem oil were exported in 1990 valued at 300,000 rupees.)

To increase the amount of oil harvested, the Khadi and Village Industries Commission has pioneered various aspects of processing the fruit and seeds over the past two decades. This grass-roots organization located in Pune has been the leading advocate for neem oil as a resource for India's villagers. Already, it has created the makings of a major village industry, developed on a rational and organized basis.

One difficulty, as with most oilseeds of the forest, is that neem must be harvested during the wet season, and without local drying facilities the fruits and seeds rapidly deteriorate and become contaminated with aflatoxin (see page 53). Ideally, the fruits should be depulped without delay and the seeds thoroughly dried. The Khadi and Village Industries Commission has devised and popularized simple methods for depulping, drying, and decorticating neem products, even in the remotest villages.

costly solvents and complex facilities. So far, at least, little oil has been produced this way.

Some of the many everyday uses for neem oil in India are discussed below.

Soap

India's supply of neem oil is now used mostly by soap manufacturers. Although much of it goes to small-scale specialty soaps, large-scale producers also use it, mainly because it is cheap. Generally, the crude oil is used to produce coarse laundry soaps. However, more expensive soaps are made by saponifying the crude oil and distilling the resulting fatty acids before adding the lye. The resulting almost colorless and odorless product is suitable for top-quality toilet and laundry soaps.[3]

Cosmetics

Neem is perceived in India as a beauty aid. Powdered leaves, for example, are a major component of at least one widely used facial cream. Purified neem oil is also used in nail polish and other cosmetics.

Lubricants

Neem oil is nondrying, and it resists degradation better than most vegetable oils. In rural India it is commonly used to grease cart wheels. It could find many similar lubrication applications in other locations, especially in village settings in the warmer parts of the world where neem can be grown.

FERTILIZERS

Neem has demonstrated considerable potential as a fertilizer. For this purpose, neem cake and neem leaves are especially promising.

Neem Cake

The residue left after the oil has been removed varies widely in composition. However, the broad ranges in composition are:

Crude protein	13–35 percent
Carbohydrate	26–50 percent
Crude fiber	8–26 percent
Fat	2–13 percent
Ash	5–18 percent
Acid-insoluble ash	1–17 percent

This so-called "neem cake" has considerable local potential. Although too bitter for animal feed, it seems to have unique promise as a fertilizer. It contains more nitrogen, phosphorus, potassium, calcium, and magnesium than farmyard manure or sewage sludge.[4] It is widely used in India to fertilize cash crops, particularly sugarcane and vegetables. Plowed into the soil, it protects plant roots from nematodes and white ants, probably due to its content of the residual limonoids.

[3] Some are being exported. Neem soaps, toothpastes, oil, and leaves are widely available in Canada and Britain, for example.
[4] Radwanski and Wickens, 1981.

Accra plains, Ghana. Stacks of neem firewood for sale are a common sight beside many roads in West Africa. (E.S. Ayensu)

Surprisingly, neem cake sometimes seems to make soil more fertile than calculations predict.[5] This is apparently due to an ingredient that blocks soil bacteria from converting nitrogenous compounds into (useless) nitrogen gas. When mixed with urea, for example, neem cake cuts down on the amount of urea converted to nitrogen gas in the soil. So far, this finding, which might prove to be a major breakthrough, has not been pursued beyond the laboratory. If it proves real in everyday practice, it might boost the effectiveness of fertilizers everywhere—restoring to the soil that part of their power now lost by bacterial action.

Neem Leaves

The cake is not the only source of fertilizer. In some areas of India's Karnataka State, people grow the tree mainly for its green leaves and twigs, which they "puddle" into flooded rice fields before the rice seedlings are transplanted.

---

[5] This feature has been measured especially when added to the water in rice fields.

Neem leaves have been used as mulch in tobacco fields in the Jaffna district of Sri Lanka.[6] In The Gambia, tomato plants matured several weeks earlier and had more numerous and longer branches when mulched with neem leaves.[7]

TIMBER

As noted previously, neem is a member of the mahogany family, and the properties of its wood resemble mahogany. It is relatively heavy, with a specific gravity varying from 0.56 to 0.85 (average, 0.68). When freshly cut, it has a strong smell. Although easily sawn, worked, polished, and glued, it must be dried carefully because it often splits and warps. It also splits easily when nailed, so that holes must be prebored. Nevertheless, it is a good construction timber and is widely used in carts, tool handles, and agricultural implements. In South India it is a common furniture wood.

The heartwood is red when first exposed, but in sunlight it fades to reddish brown. It is aromatic, beautifully mottled, narrowly interlocked, and medium to coarse in texture. It is subject to only slight shrinkage and can be readily worked by hand or machine. Although it lends itself to carving, it does not take a high polish.

The timber is durable even in exposed situations. It is seldom attacked by termites, is resistant to woodworms, and it makes useful fence posts and poles for house construction. Pole wood is especially important in developing countries; the tree's ability to resprout after cutting and to regrow its canopy after pollarding makes neem highly suited to pole production.

FUEL

Neem produces several useful fuels. As mentioned above, its oil is burned in lamps throughout India. In addition, its wood has long been used for firewood. Moreover, the husk from the seeds—containing no oil and representing the bulk of the wastage in pesticide manufacture—is mainly employed as fuel.

Because of the tree's good growth and valuable firewood, it has become the most important plantation species in northern Nigeria. It is also grown for fuel around large towns. Charcoal made from this neem wood is of excellent quality, with a calorific value only slightly below that of coal from Nigeria's Enugu mines.

---

[6] H.F. Macmillan. 1962. *Tropical Planting and Gardening*. Macmillan, London.
[7] Redknap, 1981.

OTHER PRODUCTS

Several products in addition to those previously discussed have been generated from neem. Among them are the following examples.

- **Resin** An exudate can be "tapped" from the trunk by wounding the bark. This high-protein material is not a substitute for polysaccharide gums, such as gum arabic. It may, however, have a potential as a food additive, and it is widely used in South Asia as "neem glue."[8]
- **Bark** Neem bark contains 14 percent tannins, an amount similar to that in conventional tannin-yielding trees (such as *Acacia decurrens*). Moreover, it yields a strong, coarse fiber commonly woven into ropes in the villages of India.
- **Honey** In parts of Asia neem honey commands premium prices, and people promote apiculture by planting neem trees.
- **Food** There are odd reports of people eating neem. Leaf teas may be harmful, especially if drunk in quantity over a long period, but it is said that Mahatma Gandhi, who had a hearty respect for the nutritive value of greens, commonly prepared a neem-leaf chutney and ate it with gusto—despite its incredibly bitter taste. Recently, the discovery of a rare neem tree with "sweet" leaves has been reported.[9]
- **Fruit Pulp** Pericarp represents about half the weight of neem fruits, and when they are processed to obtain the seeds, large quantities of pulp are also produced. This neem-fruit pulp is a promising substrate for generating methane gas, and it may also serve as a carbohydrate-rich base for other industrial fermentations.[10]

---

[8] Anderson and Hendrie, 1971; Anderson et al., 1972.
[9] Patrao, 1985.
[10] Mitra, 1963.

9

Reforestation

Neem appears to be a good candidate for planting throughout most of the warmer parts of the world. It not only grows vigorously in many types of semiarid and tropical sites, it is a multipurpose species that provides villagers with various products from which to derive an income during the years when the trees are maturing. This feature is important for motivating enthusiastic local tree planting.

Already neem is regarded as a valuable forestry species in both India and parts of Africa, but even there it could become more widely employed. It is also promising for planting in areas now suffering desperate fuelwood shortages. It is useful as a windbreak, exceptional as a city tree, and it can grow in (and perhaps neutralize) acid soils that plague much of the tropics.

This evergreen, which sheds its leaves only under extreme heat and drought (and then only for a short time), is valued for its shade. Its extended branches make it excellent for parks, roadsides, villages, streets, courtyards, shelterbelts—in fact, almost any place where some relief from the sun would be appreciated.

For these and other purposes, the tree is likely to be popular. It is impressive to see how it has been accepted, even eagerly sought, by people in the Sahel.[1] To a large extent it is being spread there through private and commercial initiatives. Even so, the Sahel uses the tree far less than it could.

Planting neem on a large scale might also improve the declining ecosystems of many areas considered fairly hopeless. In Haiti, for example, and other countries where the tree cover has been stripped away, vast plantings of any type of tree would likely bring environmental benefits, among them fewer floods, less siltation, and reduced

---

[1] According to forester Jean Gorse, who spent most of his career working in the Sahel: "The propagation of neem in West Africa in the '50s and '60s must be considered as the most successful (and cheapest) rural forestry operation ever implemented in this region. This was mainly due to the 'attractiveness' of neem."

erosion. Neem is just one candidate for such reforestation, of course, but it is a good one.

This tree is a farmer's friend and, when people know it better, it is likely to stimulate much spontaneous planting, especially as markets for its fruits and seeds develop. On the farm and around the house neem is useful not only as a windbreak and a welcome source of shade, but its seedcake is a good fertilizer—containing (as we have noted) nitrogen, potash, phosphorus, calcium, and magnesium.

ESTABLISHMENT

Neem is usually easy to establish. It grows best on deep, well-drained sandy soils. However, it often fails on silty or micaceous loams and silty clays, in depressions with slow drainage, and in soils with high or seasonally fluctuating water tables.

In their first months after transplanting from a nursery, neem seedlings greatly benefit from tillage, weeding, irrigation, and one or two fertilizations.

Young plants develop fairly rapidly, at least after the first season. As a rule their girth increases 2–3 cm a year, although even faster growth is often attained.

Neem needs open sunlight for best performance, but seedlings vigorously push their way up through thorny scrub and even crop plants.[2] The seedlings begin by emphasizing root growth. Only when roots are well established does the overhead growth become rapid. In harsh environments and on poor soils, this early emphasis on establishing extensive roots endows the tree with exceptional ability to survive adversity.

Although neem can be raised in nurseries and transplanted as seedlings, direct sowing on the site is sometimes easier and more successful. Seeds should be taken from thoroughly ripe fruits picked directly off the trees. They should be sown as quickly as possible.

Examples of some experiences with planting neem follow.

ASIA

Neem has been planted in many parts of Asia: Bangladesh, Burma, Cambodia, India, Indonesia, Iran, Malaysia, Nepal, Pakistan, Sri

[2] Tree planters have taken advantage of this feature in trials in West Africa. They planted neem along with pearl millet, and, after the crop was harvested, a good stand of healthy young trees remained. This made possible the establishment of a neem plantation at a relatively low cost. Neem did surprisingly well under pearl millet. Although hidden beneath the towering crop (3 m tall) for several months, it did not appear to suffer.

Lanka, Thailand, and Vietnam. It has recently been introduced into Saudi Arabia, the northern plains of Yemen, and China (Hainan Island).

India

In India, neem grows wild in dry forests and is also cultivated in all but the highest, coldest parts of the country. It thrives best in the drier zones of the northwest, and a large number of trees are found in the state of Uttar Pradesh. It is commonly planted as a roadside tree to form shady avenues. Visitors to New Delhi cannot fail to admire the stately neems adorning the avenues, spreading from both sides a thick green canopy that shields people from the fierce sun.

Burma

Although Burma is one of the main countries where neem is native, not much about its neem trees has been recorded. Nonetheless, in recent years a German aid project has helped Burmese scientists develop a neem-seed pesticide. This one-step, formulated, methanolic extract is produced by a pilot factory in Mandalay. The product has become popular among local farmers, who use it for controlling vegetable and peanut pests. The factory also produces neem oil and sells it locally for manufacturing soap and candles.

Indonesia

Java has an enormous array of different types of neem trees. Some growing on a small commercial plantation have recently been found to have seeds that are extraordinarily effective against insects due to their high content of active compounds.[3]

Pakistan

Neem is fairly widespread in the country south of Lahore. In many cities giant neem trees, more than 100 years old and more than 30 m tall, grace many roads.

---

[3] Information from M. Jacobson.

Opposite: neem in Tumkur, India. India has an estimated 18 million neem trees. They make excellent avenue trees and provide both beauty and shade. Hindus regard the tree almost with reverence. They treat ailments such as tuberculosis, warts, snake bites, malaria, and diabetes with it. Each New Year, certain Hindus eat the intensely bitter leaves and bathe in water in which the leaves have been steeped. To them, this ritual cleanses the body and the soul, thereby ensuring good health in the coming year. (J. Walter)

Philippines

Neem was introduced to the Philippines only in 1978, by scientists working at the International Rice Research Institute (IRRI). By 1990, however, IRRI had distributed more that 120,000 seedlings and the tree was growing on at least eight islands. Widescale plantings for fuelwood and potential pesticide production had also been undertaken by private and governmental agencies. Owing to numerous typhoons, neem is unsuited to the northern and central regions, but in the south it grows well.

Saudi Arabia

Introduced into the country more than 40 years ago, the tree has acclimated remarkably well to the hot and arid conditions. It is probably more common than date palm or any other tree as an avenue tree and can be seen in Jeddah and other cities.

In the plains where the Prophet Muhammad is said to have delivered his farewell sermon some 1,400 years ago, a city of thousands of tents springs up each year to accommodate the pilgrims. In the area, one of the hottest on earth, there is little relief from the intense heat—but relief is on the way. What is probably the world's largest neem plantation, about 50,000 trees, has recently been planted.[4] The project is designed to provide shade to the 2 million Muslim pilgrims who camp there annually for the hajj.

Thailand

Thailand has many "Indian" neem trees (*Azadirachta indica*) as well as its own species, *Azadirachta siamensis*, which also might have promise. It is fast growing. At Ratchaburi, on arid rock outcroppings, 200,000 specimens averaged 11 m tall only 6 years after planting. They were also heavily laden with fruits.

---

[4] Ahmed et al., 1989.

Opposite: neem trees beside a church in Dakar, Senegal. In cities throughout Senegal neem is commonly planted along the roadsides. The second-largest city, Thiès, is considered one of the most attractive in Africa because its downtown area has shady streets, cool, green parks, and attractive, tree-shrouded compounds. Thiès is an ideal example of urban forestry's power to improve the quality of life in tropical cities—and 95 percent of its trees are neems. Neems along the streets are more than just ornamentals. They are an important economic resource for the poor. The government prohibits harvesting wood from such trees, but people are allowed to gather the fruits and seeds. This helps to support hundreds of families who are in desperate need. (H. Schmutterer)

AFRICA

Indian immigrants introduced neem to Mauritius and may also have taken it to continental Africa. It is now widely cultivated in Mauritania, Senegal, The Gambia, Guinea, Ivory Coast, Ghana, Burkina Faso, Mali, Benin, Niger, Nigeria, Togo, Cameroon, Chad, Ethiopia, Sudan, Somalia, Kenya, Tanzania, and Mozambique. In each case, it is found particularly in the drier, low-lying areas.

Senegal

Because of tree planting programs of the Forestry Department and of the local people, Senegal probably has more neems than any other African country. The tree dominates towns and villages all over the country. It is used for shade and for firewood, and it has very beneficial ecological consequences, including the saving of many indigenous trees that would, in its absence, have been felled for fuel.

Ghana

Neem has been growing on the plains near Ghana's capital, Accra, since the 1920s (see sidebar). The trees have naturalized, and their spread has been boosted by birds and bats that feed on the fruits and spit out the seeds while sitting in the branches. Neem is now scattered all over the area.

With their vigorous growth, the trees have become Ghana's major source of firewood. Alongside many highways and byways, it is common to see stacks of neem wood awaiting trucking to the cities.

In Accra and other centers, neem is now a common street tree and backyard shade tree. It is normally pollarded (topped) annually and the resulting branch wood hawked for fuel or building poles.

Niger

At the beginning of the nineteenth century, the Majjia Valley in central Niger was heavily wooded. But it is located in the southern Sahel, an area with highly variable and low rainfall (400–600 mm a year). The growing population—with a relentless appetite for fuelwood, fodder, and construction materials—left it bare. By the drought years of the early 1970s, wind erosion was blowing away nearly 20 tons of topsoil per hectare per year. In the rainy season, wind-blown sediment would smother farmers' seedlings, forcing them to reseed their fields over and over.

In 1975, the American relief agency CARE began planting neem

How Neem Reached Africa

To people in West Africa these days, neem seems like an established part of the countryside. The general feeling is that it has been there since time immemorial. However, this tree is actually a recent addition to the African scene.

It was Brigadier-General Sir Frederick G. Guggisberg who brought neem to Ghana, for example. He was governor (of what was then known as the Gold Coast) from 1919 to 1927, and he introduced seeds or seedlings from India sometime during that period. The first were planted in the Northern Territories. Today, as a result, neem is found throughout Ghana and the Sahelian region.

The governor's efforts have given rise to at least two local names for neem. In Ghana, the tree is normally called "king," which is the local title for governor. In Mali, the vernacular name in the Dyula language is "goo-gay," a corruption of "Guggisberg."

Neem was first introduced to Nigeria in 1928 (probably from Ghana), where it was successfully established in the Bornu province. Several thousand seedlings from the first plantation were replanted in Sokoto, Katsina, and Kano provinces in the 1930s. Neem was also successfully established by sowing fresh seed directly into the shelter of indigenous vegetation and local food crops. There are now considerable plantations for firewood and construction materials throughout those areas of northern Nigeria.

Neem also seems to be an entrenched part of the scene in The Sudan. There, it is valued mainly as a street and amenity tree and is commonly seen at railway stations and beside mosques. Here again, however, neem is a new arrival, historically speaking. The first ones apparently were planted at Shambat in 1916. Probably, they were brought directly from India by a diligent colonial forester who appreciated their value for producing shade, fuel, wood, and oil for lamps.

Just how Senegal got its first neem is uncertain, but in the 1950s Senegalese agronomist Djibril Sene went to India to gather neem seeds. Many, if not most, of the neems now seen throughout the country result from his far-sighted efforts.

windbreaks. By 1987, some 560 km of double rows of neem were established and more than 3,000 hectares of cropland protected. The trees cut wind velocity near the ground by 45–80 percent, resulting in less erosion and more soil moisture. Crop yields jumped 15 percent or more, even after accounting for the land taken up by the lines of trees.

Moreover, the obvious beneficial effect of the windbreaks—particularly as a source of cash from the sale of wood—has encouraged some farmers to start their own nurseries. Currently, more than 100 private nurseries are being tended. The long-term success of the project seems to be assured by the spread of the woodlots and nurseries into private control.

Nigeria

Neem is common, especially in towns and villages, in the northern regions. Sometimes it is planted in large numbers along roadsides—along the road between Maiduguri and Lake Chad, for instance.[5]

Mali

Neem is part of the scene along the Niger river as far north as Timbuktu. Many of the trees are pollarded (at about 2 m height) to provide forage to cattle and goats. Many are also pruned into unusual shapes by camels.

Sudan

Sudan was one of the first African countries to get neem. Today, the trees are widespread along the Blue and White Niles, in irrigation schemes, and in towns and villages.

THE AMERICAS

Apparently, it was immigrants from India who introduced neem to several Caribbean nations. The tree is now grown as a medicinal plant in Suriname, Guyana, Trinidad and Tobago, Barbados, Jamaica, and elsewhere. More recent neem plantings are also found in St. Lucia, Antigua, Dominican Republic, Mexico, Belize, Guatemala, Honduras, Nicaragua, Bolivia, Ecuador, and Brazil. In most of these nations,

[5] Information from H. Schmutterer.

however, the plantings are small, scattered, and exploratory. Only in Haiti, the Dominican Republic, and Nicaragua have large numbers of neem trees been planted so far.

Haiti

In the last decade or so, neem has been widely planted in Haiti. In fact, this tree is now one of the leading species for reforesting this much-denuded land. For example, one project funded by USAID has planted 200,000 neem trees as part of a road beautification program using seed imported from Africa in the late 1970s. Later, neem became a popular species for planting. The trees have grown so well that today neem seed is becoming a Haitian export. Approximately 40 tons were processed for azadirachtin by an American company in 1990. Since then, other companies have also sought to buy Haiti's neem seed.

United States

Because the tree is a tropical species, it probably cannot be grown economically in the continental United States beyond South Florida. In South Florida, however, there are four mature neem trees (two in Miami and two in Fort Myers) and 50 smaller ones (in Homestead). There are also eight trees in the futuristic Biosphere 2.[6]

Of course, the tree can thrive in Hawaii and other locations in the American tropics. Researchers have already begun planting it in Puerto Rico and the Virgin Islands, for example. A specimen planted at the East-West Center in Honolulu, Hawaii, in 1984 was nearly 10 m tall and fruiting heavily in 1991. And in 1989 the Hawaii State Senate passed a resolution supporting research and development of this "wonder tree."

THE PACIFIC

Nineteenth century immigrants carried the tree from India to Fiji, and it has since spread to other islands in the South Pacific, even to Easter Island, which is hardly known as a place for trees. In Papua New Guinea neem was introduced at the beginning of the 1980s, mainly in the Port Moresby area.

---

[6] This is a $150-million, 1.2-hectare set of glass and steel greenhouses in the desert near Tucson, Arizona, in which 8 humans are sealed for 2 years. The greenhouses are isolated from all inputs such as air, water, and fuel and are filled with plant and animal life, to represent a self-sustaining microcosm of the planet Earth (Biosphere 1).

10

Next Steps

If neem lives up to its early promise it will help to control many of the world's pests and diseases, as well as reduce erosion, desertification, deforestation, and perhaps even slow the rate of increase in population. So many details remain to be fleshed out, however, that its practical possibilities cannot yet be seen even to a limited extent.

The fact that neem is a tree is in some sense a limitation; to mass-produce products on a vast scale from trees is much harder than from annual plants. However, trees also have several advantages. They are perennials that will provide their products for decades, they pose little risk of becoming weeds, and, once established, they require little care. Moreover, growing tree crops these days is an advantage in itself. Indeed, resources harvested from trees are of vital importance to this seriously threatened planet. Reforestation contributes to a better world, and neem is a good candidate for global tree planting.

Whether neem will thrive in dense plantation blocks is not absolutely certain, but there are many sites where it seems ideal. It can grow in certain marginal lands, for example, and therefore does not have to displace food production because it can be raised where soils are too worn out for crops. It even benefits certain types of soils and, like all trees, helps reduce erosion.

Harvesting neem fruits does not destroy the tree; unlike most reforestation species, neem is more profitable standing than felled. Thus, the use of neem products has the merit of promoting a greening of the earth.

GENERAL ACTIONS

Despite some unresolved questions, enough is already known that exploiting certain neem uses can begin immediately. Indeed, the global problems posed by pests, diseases, erosion, deforestation, and desertification are so vast that boldness is called for and some risk is worth taking. This, therefore, is a time for people to bring the plant and its products into international use. An orderly creation of plantations and markets—with reliable availability of uniform, good-quality seeds at stable prices—could see neem rise steadily to become one of the most widely grown trees in the world—perhaps eventually rivaling the African oil palm in its value.

Governments, agencies, and individuals that assist developing nations should support the development of neem plantings, underwrite projects to harvest and process the seeds for use in pest control and personal hygiene, and assist countries to develop high-quality ecotypes in terms of azadirachtin content and other desirable traits.

In developing neem, there is potential for much innovation. One example is the concept of centering rural industries around neem-extraction facilities.[1] In this system, industrial development would be integrated with neem-tree growing. It might incorporate crops and livestock, but growing neem trees and processing their products would form the core. This integrated combination has a good chance of providing sustainable, self-reliant, and decentralized rural development—a long-sought goal of many economic development programs.[2] In addition, it could help national interests by reducing pesticide imports and perhaps increasing exports.

PESTICIDE RESEARCH

In the coming years, the struggle to keep food out of the jaws of plant-eating pests will increase in importance as human populations increase, living standards rise, demands for quality food (and the consequent emphasis on blemish-free fruits and vegetables) increase, and the public clamor to eliminate synthetic insecticides becomes more insistent. Neem could be the key to opening this new era of safer pest-control products and, if so, is likely to be in huge demand.

Although research on neem-based pesticides is under way, it is only a fraction of what it might be. Currently, there are projects in Australia, Bangladesh, Burma, Canada, Dominican Republic, Germany, India, Israel, Kenya, Nigeria, Niger, Mali, Pakistan, and the United States.[3] Nonetheless, most are small, undersupported, and tacked onto other

---

[1] This idea was first proposed in Michel-Kim and Brandt, 1981, as well as by S.A. Radwanski.

[2] Something similar has been created in Taiwan, where industrial parks are centered around areas of bamboo production.

[3] Most of this work has been summarized at three international conferences. (See Schmutterer et al., 1981; Schmutterer and Ascher, 1984; and Schmutterer and Ascher, 1987.)

projects. Much greater effort is warranted, and some of the topics for research are discussed below.

Preparation of Extracts

There is now enough information to encourage use of the current formulations. However, "low-tech" methods should be devised to extract and formulate neem materials in ways that can easily be undertaken by farmers who grow their own neem trees for their own use.

At present, the quality of neem extracts varies. The differences seem to depend on the way the seed was handled, stored, or extracted, and perhaps other factors yet to be recognized. Thus, research on the optimal handling of neem materials is particularly needed. Topics to be studied include conditions for storing seed before extraction and before use, as well as the effects of storing and handling the extracts after they have been made. Increasing the storage life of neem formulations is vital.

Before mass-producing neem products for international use, standardization is essential. The efficacy of various batches is impossible to compare now because no standard of potency is being used. An international nomenclature for neem ingredients—perhaps defined in ppm of azadirachtin—would help bring some order out of the current chaos.[4]

Studies of Effectiveness

The potential of neem products as a village-level remedy for the agricultural pests of the tropics should be fully explored. Further research on specific crops and sites, pest organisms, formulations, and application methods is needed.

Modes of Action

Neem derivatives are promising pest control materials, but just how they work on various species is a topic deserving much greater research attention. Basic research to study the effects of neem extracts on hormone regulation and hormone receptors is required.

Formulation of Products

Additional research is also needed to extend the period in which neem products remain active. When sprayed on plants, the extracts

[4] Methods are established and are being used for the quality control of Margosan-O®, but so far at least, no one else uses the same quality-control methods.

Neem's Predecessor

The public's increasing concern for the environment seems likely to result in a rising demand for pesticides from plants rather than from petroleum. Such "soft" pesticides represent the hope that agricultural pests can be controlled while maintaining environmental stability. One American market-research firm estimates that by 1998, total soft-pesticide sales in the United States alone will reach $813 million annually, up from $450 million at present.

Neem may become a major part of that growth, but it is not the first botanical pesticide. Pyrethrins, which are naturally derived from daisylike flowers of certain species of *Chrysanthemum*, have been used for centuries. Almost 2,000 years ago the Chinese knew that chrysanthemum plants had insecticidal value; some 2,400 years ago the Persians used them. Not until recent centuries, however, were the potentials of the pyrethrins, extracted from the flowers, fully appreciated. Supposedly an Armenian trader, who had learned the secret while traveling in the Caucasus, introduced the insecticide into Europe early in the nineteenth century. Last century, Dalmatia (Yugoslavia) became the center of the world's pyrethrum industry, but after World War I, Japan became the main producer. With supplies cut off during World War II, the Allies began producing the flowers in Kenya. Since the 1960s pyrethrum production has been established in the New Guinea highlands also.

Like neem products, pyrethrins are valued for their low toxicity to mammals and birds. However, the ingredients in these insecticidal chrysanthemums are lethal to insects in a different way from those in neem. They are nerve poisons and contact insecticides. Pyrethrum has quick knockdown properties and is the active ingredient in millions of aerosol spray cans people use against flies and mosquitoes.

Despite the development of many synthetic insecticides, this chemical from chrysanthemums has maintained its position as a major commercial product. World production is more than 10,000 tons. Although powerful synthetic analogues have been developed, demand for the natural material has remained high, and in the past several years it has been in short supply.

Now neem, another botanical pesticide, can perhaps step up to take an equally important, but complementary, role in the rising soft pesticide market.

are degraded by sunlight within days. As previously noted, one American company has found that retaining a portion of the seed oil and adding an ultraviolet screen extends the activity to several weeks. This, however, relies on industrial ingredients and is covered by a patent.[5] Ways to achieve similar effects in Third World settings are now necessary.

Safety

The product Margosan-O® (see Appendix A) has been tested and certified safe (when used as directed), but more toxicological research on neem extracts is needed. Chronic-exposure tests, higher-mammal studies, and epidemiological evaluations could help identify any potential short- and long-term hazards before massive international use. All in all, appropriate researchers should undertake studies to assess any remaining possibility of toxicity to higher mammals, birds, or fish.

Pest Resistance

Because of the complexity of the mixtures and their modes of action, it seems unlikely that any resistance to mixtures of neem products will develop in the short run. However, insects have disproved similar projections with previous pesticides too often for complacency. Neem materials should therefore be used circumspectly. If applied by judicious spot treatments at appropriate times, they may remain effective for centuries. On the other hand, if used indiscriminately in blanket sprays, they may induce resistance in the pests and be rendered ineffective within a few years.

The buildup of resistance is much more likely with refined neem formulations based on a single active ingredient from neem, such as azadirachtin. Pests can probably develop resistance to a single neem ingredient about as readily as to other insecticidal compounds.

Further research into the issue of resistance is called for.

Human Resistance

Recent surveys reveal that in both India and Pakistan most of the poorer farmers mix a "handful" of neem leaves in their stored grains to protect them from pests. However, the more affluent farmers, although aware of this practice, do not follow it. Some questioned its efficacy, but most did not want to be stigmatized as "backward" for

[5] Larson, 1987.

following an ancient and traditional practice.[6] The key to quickly overcoming this misguided attitude is to show that using neem is actually more modern than the modern techniques for which these farmers are paying big money.

Specialized Uses

Many specialized insecticide uses deserve research, especially in tropical areas. One example is neem-based insect-repelling treatments for common products, such as the bags used for holding and shipping food and other perishables.

STRUCTURAL ANALYSIS

Neem materials are a vast storehouse of possible pest-control agents of the future. Azadirachtin, meliantriol, and salannin, for instance, might serve as models for the synthesis of insect-feeding inhibitors and growth regulators for controlling stored-grain pests, grasshoppers, locusts, nematodes, and other pests. Even if such synthetic analogues prove commercially feasible, it is unlikely they will cut greatly into the markets for the directly extracted neem materials.

NEEM OIL RESEARCH

Despite centuries of use in India, neem oil is still poorly understood as compared to palm oil, soybean, and other vegetable oils. Some basic chemistry, as well as processing and product-development research should be most useful.

Oil Purification

The methods used for processing and refining neem fruits and seeds all need improvement. In particular, simple methods that farmers can employ themselves are comparatively inefficient at present. On the other hand, research on the use of advanced separation technology is also required. Problems of deodorizing, refining, and purifying the oil in industrial production have yet to be made practical and economic on a large scale. Modern separation processes, such as selective adsorption or high-tech membranes, might prove extremely valuable here.

---

[6] Ahmed, 1990.

Neem and the Superbug

As we go to press in December 1991, news is sweeping the nation that a deadly insect infestation has destroyed America's winter melon crop and damaged its lettuce, cabbage, broccoli, cauliflower, and carrot crops. Millions of voracious insects have spread over California's Imperial Valley, massing on the undersides of leaves and sucking plants dry, weakening or killing them in the process. American consumers have been told to expect serious shortages of some fruits and vegetables, not to mention soaring prices.

The poinsettia whitefly, or "superbug" as farmers are calling it, is a new, more potent strain of the sweet potato whitefly (*Bemisia tabaci*). It appears to be pesticide resistant and eats "just about everything" in its path. According to California agriculture experts, asparagus and onions are the only crops that it does not like.

As a result of the tiny fly's attack, California's governor, Pete Wilson, has declared a state of emergency. And no wonder. California farmers have suffered nearly $90 million in damage, more than 2,500 farm workers are out of work, and hundreds of farm-related businesses have had to take huge losses. California's agriculture experts expect that eventually the blizzard-like swarms of tiny flies will destroy $200 million worth of winter vegetables.

What will happen in future years is anybody's guess. California has no native predators that are effective against the superbug, and all the authorized pesticides are largely useless. However, neem is one of the possible answers to the problem. For several years, this very same insect has been one of the prime targets of neem-seed extracts. This was in other parts of the country and on other crops—mainly ornamental plants and mainly in greenhouses. There, neem products have controlled the superbug very effectively, but whether they will be the answer to the problem over the vast areas of vegetables and fruits growing in the Imperial Valley is as yet uncertain.

One major problem is that neem is not registered for use on food crops. Another is the lack of supplies of neem seeds. Nonetheless, even the possibility of a natural pesticide for such a knotty problem is cause for hope that a cure can be found.

Product Development

Neem soaps, lubricants, and many other consumer products offer exciting promise, especially for tropical countries. Here, too, there is much scope for invention and product development. Basic needs include formulations, analyses, and standards for quality.

NONINSECT PESTS

Neem opens up many possibilities of new products that could benefit horticulture, silviculture, and agriculture. There is much scope for research and development in this area. It is perhaps not too far-fetched to speculate that the tree's extracts might be employed in the following ways:

- As systemic fungicides for treating sick trees or crops;
- For preventatives that could stop fungal diseases from establishing themselves in plants;
- In treatments for trees diseased by viruses; and
- In treatments for crops threatened by garden snails and slugs.

HUMAN HEALTH

Studies of neem's medicinal values are urgently needed and include the following topics:

- Effectiveness in alleviating pain or fever;
- Antibacterial and antiviral qualities;
- Control of dental cavities and pyorrhea;
- Use of neem twigs for teeth cleaning in areas where toothpastes are unavailable or beyond the budgets of poor people;
- Topical treatments for lice;
- Use of neem-leaf juice and neem oil in the treatment of psoriasis;
- Topical treatment for warts;
- Treatment for parasites in the human digestive tract; and
- Treatment for parasites in the human blood and lymph systems, including those causing malaria, Chagas' disease, river blindness, elephantiasis, and sleeping sickness.

VETERINARY MEDICINE

Under normal use, neem apparently affects a variety of organisms, including bacteria, fungi, mollusks, and protozoan parasites, which may open many avenues for exploratory research in veterinary medicine. Traditionally, Indians have rubbed neem products onto livestock to treat various complaints. Research should be undertaken to confirm the ability of neem oil or neem-seed extracts or a combination of both to repel insects and ticks, as well as to soothe cuts and bruises and to cure scabies. Neem may also help with several serious tropical skin parasites—those that cause mange in camels and donkeys, for example.

Neem products should also be tested as a treatment for intestinal parasites, such as roundworms and tapeworms. The oil's efficacy in the treatment of infections, particularly of the genital tract (postpartum inflammations of the uterus, for instance) in animals also deserves attention.

GENETIC IMPROVEMENT

Individual neem trees vary greatly in their morphology and perhaps in their chemical makeup. It is not yet understood whether these differences are based on genetics or environment or both, although it is believed that environmental factors (such as drought stress) play a dominant role. Basic research is needed in this area and will have to be carried out mainly in Asia, where the greatest range of genotypes is to be found.

So far there has been no selection or breeding for maximum pesticide production. One approach is to seek out the trees whose seeds have the highest proportion of azadirachtin. This would have to be done using standardized methods to ensure that differences in seed handling, deterioration, and other features do not interfere. A major breakthrough would arise here if the azadirachtin content can be correlated with a visual or readily identified feature of the trees or seedlings. With millions of neems in the world, a rapid qualitative assessment would be most valuable at this time.

The second approach is to select trees that yield maximum numbers of large fruits. The number and weights of fruits on different trees vary greatly, and obtaining the maximum yield of kernels may be economically more important than the percent of azadirachtin in each kernel.

BIOTECHNOLOGY

Biotechnology research that might benefit neem includes:

- Magnifying desired traits;
- Examining the enzymology and gene expression of limonoid production;

- Transferring neem's pest-resistance genes into agriculturally significant plants; and
- Genetically mapping neem's DNA. (This will speed the development of neem products and benefits.)

REFORESTATION

Much valuable research could be done in the area of neem silviculture. This might include assessments of the following:

- Taproot effects.
- Lateral (feeding) root effects.
- Mycorrhizae.
- Other beneficial soil microbes.
- Seed viability. (Research is particularly needed to develop methods to extend the period of the viability of neem seeds for replanting.)
- Rapid establishment. (At present, the horticultural conditions and practices that lead to optimal growth or production are unknown. Needed are ways to speed up establishment of the trees.)
- Provenance selection (selection of genotypes better suited for select sites—relatively dry areas, for instance).

In addition to such silvicultural studies, neem's apparently remarkable ability to survive in cities and to withstand excessive heat, as well as survive air and water pollution should be evaluated. This could well boost its use in urban forestry throughout the tropics.

Wood Products

The possibility of selecting genotypes for the production of various types of wood products should also be examined.

One need is for types with straight trunks and a maximum length of clean bole. These would be used to produce construction lumber.

Another need is for easy-pollarding types suitable for producing building poles. In several countries neem poles are more valued than any other neem products. Genetic selection for optimum branching from a stump cut close to the ground could be helpful here, as could research to determine the best time of the year and the best height at which to cut the trees. Rural producers in Burkina Faso already manage their neems like hedges (*tenkodogo*) to harvest building poles more easily.

Fruit Orchards

Frequent coppicing or pollarding are not conducive to good flowering because they severely restrict the growth of lateral (flower-bearing) branches. This, in turn, reduces the production of fruit. Therefore, to grow neem for its seed and oil requires a different approach.

A neem-fruit plantation of the future will likely consist of trees specially selected for high yields of high oil-bearing seeds and widely spaced to allow for an optimum spread of the lateral branches and an unrestricted formation of the flowers. Producing and managing these plantations has little in common with conventional forestry. Indeed, it is more in the domain of pomology: the art and science of cultivating fruit trees. Sophisticated modern techniques such as clonal selection, tissue culture, pruning, grafting, mulching, and fertilizing with major, minor, and trace elements are all research requirements.

Clarification of such features is important. Neem orchards established in this way could provide a regular supply of quality seed and oil—and thereby become the basis for a thriving international industry based on neem ingredients.

Agroforestry

By and large, neem appears to be a poor companion for field crops. However, certain plantings might prove suitable for integrating with local farming and grazing practices. Further investigation should be made. Farming systems combining neem, fast-growing timber trees, and shrubs as a combination fallow would likely turn around declining ecosystems of many humid and semiarid tropics while providing a continuing income. Such systems may help restore and maintain soil fertility.

Plant Viruses

In certain old (1920s) experiments it was reported that neem-leaf extracts seemed to overcome viral diseases in beans, tobacco, and some other crops. This could prove to be of outstanding importance. On the other hand, the results were inconsistent and there is likelihood that they will prove unrepeatable.

Nonetheless, if neem shows even limited antiviral activity, that alone would be of interest to world agriculture. The fact that its compounds are systemic and that the tree grows in many countries where viruses devastate crops (streak virus in Africa's corn is an example) are additional benefits of possibly enormous consequence. This is a shot in the dark, but worth exploring.

Plant Bacteria

Asians have long used neem to treat bacterial diseases of the skin (see Chapter 7), but, at least for now, its use to combat bacterial diseases of plants is a research area wide open for exploration.

RELATED SPECIES

Neem (*Azadirachta indica*) has at least two close relatives, *A. siamensis* and *A. excelsa*. They, too, are promising resources.

A. siamensis is known as "edible neem" because its young leaves and flowers contain lower amounts of bitter principles than *A. indica* and are consumed in considerable quantities as a vegetable by people in Burma and Thailand. No negative consequences have been reported. *A. siamensis* also contains azadirachtin in its seed kernels and might be a useful source of pest-control materials as well as food.[7]

A. excelsa is a little-known tree of Southeast Asia. Recently, German researchers have isolated and characterized a new limonoid from its seed kernels. This compound, marrangin, shows the same mode of action as azadirachtin but is two to three times more active. Leaf extracts of *A. excelsa* also show a better efficacy than those of neem itself.[8]

[7] Information from H. Schmutterer.
[8] Information from H. Schmutterer.

APPENDIX A

Safety Tests

In 1985, as was mentioned in the introduction, the Environmental Protection Agency approved a commercial neem-based insecticide for certain nonfood uses. Called Margosan-O®, the product is currently available in limited quantities, but demand for it is said to be high and its use is increasing quickly.

Currently, Margosan-O® is registered for control of whiteflies, thrips, mealy bug, leafminers, looper, caterpillars, beet armyworms, aphids, ants, flies, cockroaches, fleas, weevils, psyllids, webworms, hornworms, spruce budworms, pin sawflies, and gypsy moths in greenhouses, commercial nurseries, forests, and homes. This is based on the results of toxicity studies required by the Environmental Protection Agency (EPA).[1]

Margosan-O® is an ethanolic neem extract concentrate having 3,000 ppm azadirachtin (± 10 percent) and is based on the original process developed by the U.S. Department of Agriculture (USDA) in Beltsville, Maryland. In a collaborative effort, Vikwood, Ltd. of Sheboygan, Wisconsin, undertook the challenge to stabilize and enhance the extract and attempt to bring it to the commercial market. In pilot plant runs, stability has been achieved up to 1 year under ambient conditions, and in excess of 3 years under refrigeration. Efficacy tests run by the USDA-Beltsville, Maryland, show activity in excess of 21 days. EPA registration has been granted for use on nonfood crops, and a U.S. patent has been granted on the product.

Following are the toxicity tests ordered by the EPA to gain registration for Margosan-O® to be used on nonfood crops.

Test 1—Avian Single-Dose Oral LD_{50}. Margosan-O® was administered to mallard ducks in order to determine a dose lethal to 50

[1] Information in this appendix courtesy Robert Larson.

percent of the duck population. Dose levels of Margosan-O® ranged at 1–16 ml/kg of body weight. Observations showed no negative effects and all ducks remained active and healthy throughout the 14-day experimental period. The acute LD_{50} of Margosan-O® to mallard ducks is in excess of 16.0 ml/kg.

Test 2—Avian Dietary LC_{50} (lethal concentration) with bobwhite quail. The birds were given their basal diet, with additions of Margosan-O® ranging from 1,000 to 7,000 ppm. Observations showed no negative effects and the quail were active and healthy throughout the 5-day test period and 3-day recovery phase. The acute oral LC_{50} of the Margosan-O® to bobwhite quail is therefore in excess of 7,000 ppm.

Test 3—Avian Dietary LC_{50} Study with mallard ducks. The ducks were given a basal diet, plus Margosan-O® ranging from 1,000 to 7,000 ppm concentrate for 5 days. No mortalities resulted. The ducks were active and healthy throughout the test and recovery phases. The acute LC_{50} of the test material to mallard ducks is therefore in excess of 7,000 ppm.

Test 4—(No. 1), Acute Toxicity of Margosan-O® to rainbow trout. This test involved a 96-hour LC_{50} of Margosan-O® at various concentrations. The LC_{50} was 8.8 ml of Margosan-O® per liter of water, and the 96-hour no-observed-effect concentration was 5 mg/l.

Test 4—(No. 2), Acute Toxicity of Margosan-O® to bluegill sunfish. The results showed a 96-hour LC_{50} of 37 mg/l and a no-observed-effect 96-hour level of 20 mg/l. In the fish bioassay tests with trout and sunfish there were behavioral responses in static water that could probably not occur in moving water.

Test 5—Acute Toxicity of Margosan-O® to *Daphnia magna*, the water flea. The test was done on newly molted instars less than 20 hours old which were placed in a fresh aquatic habitat for up to 48 hours. The LC_{50} was 13 mg/l and the no-observed-effect concentration at 48 hours was less than 10 mg/l. Therefore, the toxicity value obtained was well within the expected range, but it indicates that Margosan-O® will affect primitive aquatic invertebrates under static conditions.

Test 6—Acute Oral Toxicity. Rats were dosed once and then observed for 14 days for abnormal behavior or mortality. No negative effects were observed and the acute oral toxicity of the test material was in excess of 5 ml/kg, the limit of the required test.

Test 7—(No. 1), Acute Dermal Toxicity. A nonpermeable patch containing 2 ml/kg body weight of Margosan-O® was placed over small shaved areas on a group of albino rabbits. No mortality resulted and acute dermal toxicity (LC_{50}) of Margosan-O® was in excess of 2 ml/kg.

Test 7—(No. 2), Primary Skin Irritation. Albino rabbits were treated with Margosan-O® applied under patches on shaved areas and on abraded areas. The results showed low to moderate primary irritation to the shaved area patch and high to moderate irritation to the abraded area.

Test 8—Acute Inhalation Study. Albino rats were exposed to a total of 15.8 g of test material (estimated concentration of 43.9 mg/l/hr) for 4 hours. (This test was recently repeated and reported in terminology more acceptable to the EPA.) The LC_{50} for the Margosan-O® in the inhalation test was in excess of 43.9 mg/l/hr, the limit of the test.

Test 9—Modified Eye Irritation. Margosan-O® was administered to one washed and one unwashed eye of albino rabbits. Over 7 days both eyes showed minimal irritation.

Test 10—Immune Response. The effect of Margosan-O® on the hematology and serum electrophoretic pattern of rats, strain Sprague-Dawley, was determined. Eight male and eight female rats, each weighing between 200 and 250 g, were anesthetized by means of CO_2 and weighed. A 3-ml sample of blood was taken via cardiac puncture and the blood studied. Five male and five female rats received 0.5 ml of Margosan-O® by intraperitoneal injection. The remaining six (control) rats were left untreated. The rats were maintained until the 14th day after substance administration. At that time, they were again anesthetized and weighed and a blood sample was taken as on day 0. Blood samples were submitted to repeat the analyses conducted at the study initiation, that is, complete blood counts with differential and serum protein electrophoresis.

Body weights on day 0 and day 14 were combined for each of the four groups and a mean and standard deviation were calculated. All surviving rats gained weight and appeared active and healthy. No differences were evident between test and control animals. There were no significant changes in the hematology of the treated rats between day 0 and day 14, although statistical analysis of the electrophoretic pattern showed differences ($P<0.05$) in the globulin fractions during that period. The differential count showed a statistically significant change in the polymorphonuclear count, but none of the other differential counts differed statistically.

Comparison of the changes in blood values of the control rats over 14 days with those in the treated rats did not uncover significant differences among any of the parameters measured.

This suggests that the changes noted above were not treatment-related, but rather that they were normal and not unusual. The results of this study suggest that the test material does not cause an adverse immune response.

Test 11—Sensitization. This test was done on guinea pigs, which were shaved and patched with Margosan-O® test material for 6 hours. The procedure was repeated on alternate days for a total of nine applications. A retest dose was applied after 14 days with duplicate patches, and the reaction was read 24 hours later. Margosan-O® does not produce sensitization.

Test 12—Mutagenicity. This is the traditional Ames mutagenicity study used in the United States on five strains of *Salmonella typhimurium*. Results of this test indicate that the Margosan-O® concentrate is nonmutagenic.

Test 13—Bee Adult Toxicity Test. This test was done voluntarily and was not ordered by the EPA. With the assistance of the University of California's Apiary at Riverside, California, Vikwood, Ltd. ordered a "Bee Adult Toxicity Test" on honeybee worker adults. Margosan-O® was administered as a direct contact chemical using field dosages up to 4,478 ppm A.I./ha. It was found to be benign to honeybees at well above the recommended dosage of 20 ppm (diluted, as a foliar spray) for a common pest, the gypsy moth, *Lymantria dispar* (L.).

APPENDIX B

BREAKTHROUGHS IN POPULATION CONTROL?

Most observers conclude that today's skyrocketing growth in human population is creating a serious underlying threat to the well-being of the world's natural and economic resources. Between 1990 and 2025, for instance, the number of people on the planet is expected to rise by 3.2 billion.[1] And by the year 2000, it is projected that 1.7 billion people will be added to countries that can barely support their populations now, let alone create an economic climate to raise everyone out of poverty.

Whether neem can help reduce runaway population growth is uncertain. However, as noted earlier, exploratory research has indicated that certain neem ingredients have contraceptive properties. Thus it is possible that, given research attention, products from this tree could come into widespread use for the reduction of unwanted pregnancies. This could be an important breakthrough because perhaps 2 billion of the projected population increase before 2025 will occur in regions where neem can be grown.

Eventually, contraception could perhaps become the greatest of neem's contributions. Because of the importance of any breakthrough in this area, we reproduce in this appendix the abstracts of three recent research papers that point toward the possibility of neem providing cheap, widely available contraceptives for even the remotest regions of many of the already overpopulated nations.[2]

---

[1] United Nations Department of International Economic and Social Affairs. 1989. *World Population Prospects, 1988*. Population Studies No. 106. United Nations, New York.

[2] Other antifertility effects of neem are summarized in Jacobson, 1988, pages 145–147.

NEEM OIL FOR THE "MORNING AFTER"

Shakti N. Upadhyay, Charu Kaushic, and G.P. Talwar. 1990. Antifertility effects of neem (*Azadirachta indica*) oil by single intrauterine administration: a novel method for contraception. *Proceedings of the Royal Society of London Series B: Biological Sciences* 242:175–179.

A novel use of neem (*Azadirachta indica*) oil, a traditional plant product, for long-term and reversible blocking of fertility after a single intrauterine application is described. Female Wistar rats of proven fertility were given a single dose (100 μl) of neem oil by intrauterine route; control animals received the same volume of peanut oil. Whereas all control animals became pregnant and delivered normal litters, the rats treated with neem oil remained infertile for variable periods ranging from 107 to 180 days even after repeated matings with males of proven fertility. The block in fertility was, however, reversible, as half of the animals regained fertility and delivered normal litters by five months after treatment, without any apparent teratogenic effects. Unilateral administration of neem oil in the uterus blocked pregnancy only on the side of application, whereas the contralateral uterine horn treated with peanut oil had normally developing foetuses; no sign of implantation or foetal resorption was noted in the neem-oil-treated horn. The ovaries on both sides had 4–6 corpora lutea, indicating no effect of treatment on ovarian functions. The animals treated with neem oil showed a significant leukocytic infiltration in the uterine epithelium between days 3 and 5 post coitum, i.e., during the pre-implantation period. Intrauterine application of neem oil appears to induce a pre-implantation block in fertility; the possible mechanisms of the antifertility action are discussed.

NEEM OIL AS SPERMICIDE

Gp. Capt. K.C. Sinha and Lt. Col. S.S. Riar. 1985. Neem oil—an ideal contraceptive. *Biological Memoirs* 10(1 and 2):107–114.

Neem oil in vitro proved to be a strong spermicidal agent. Rhesus monkey and human spermatozoa became totally immotile within 30 seconds of contact with the undiluted oil.

In vivo studies in rats (20), rabbits (8), rhesus monkeys (14), and human volunteers (10) proved that neem oil applied intravaginally before sexual intercourse prevented pregnancy in all the species.

Neem oil has also been found to have anti-implantation/abortifacient effect in rats and rabbits if applied intravaginally on day 2 to day 7 of expected pregnancy. The minimum effective dose is 25 μl for rats. One month after the stoppage of neem oil application there was complete reversibility in fertility in these animals. It had no deleterious effect on the subsequent pregnancies and the offsprings.

Histopathological studies on rats' vagina, cervix, and uterus showed no ill effects of neem oil in these tissues. In contrast, nonyl-phenoxy polyethoxy ethanol, a popular vaginal contraceptive cream, showed signs of severe irritant reaction in these tissues. Radioisotope studies indicated that neem oil was not absorbed from the vagina; it thus ruled out its possible systemic effects.

Results of the present study indicate than neem oil is an "ideal" female contraceptive, being easily available, cheap, and nontoxic. Therefore, its mass acceptance is anticipated.

NEEM-LEAF EXTRACT TO REDUCE MALE FERTILITY

N.L. Sadre, Vibhavari Y. Deshpande, K.N. Mendulkar, and D.H. Nandal. 1983. Male antifertility activity of *Azadirachta indica* in different species. Pages 473–482 in H. Schmutterer and K.R.S. Ascher, eds. 1984. *Natural Pesticides from the Neem Tree* (Azadirachta indica *A. Juss.) and other tropical plants*. Deutsche Gesellschaft für Technische Zusammenarbeit (GTZ), Eschborn, Germany.

Male antifertility activity of neem (*Azadirachta indica* A. Juss) was studied in mice, rats, rabbits, and guinea pigs by daily oral feeding of a cold-water extract of fresh green neem leaves. The infertility effect was seen in treated male rats, as there was a 66.7 percent reduction in fertility after 6 weeks, 80 percent after 9 weeks, and 100 percent after 11 weeks. There was no inhibition of spermatogenesis. During this period there was no decrease in body weight and no other manifestation of toxicity were observed. There was a marked decrease in the motility of spermatozoa. The infertility in rats was not associated with loss of libido or with impotence and the animals maintained normal mating behavior. The male antifertility activity was reversible in 4 to 6 weeks and the active principle from the extract was observed to be thermostable. Neem extract also shows reversible male antifertility activity in mice without inhibition of spermatogenesis. In guinea pigs and rabbits, however, it exhibited toxicity, as demonstrated by 66.6 percent and 74.9 percent mortality in guinea pigs and 80 percent and 90 percent mortality in rabbits at the end of 4 and 6 weeks, respectively.

APPENDIX C

REFERENCES AND SELECTED READINGS

By 1991 the scientific literature dealing specifically with neem and its products had reached more than 1400 articles. In this appendix we highlight just a few of the more general reviews as well as the articles referred to in each chapter.

CONFERENCE REPORTS

Three international neem conferences have been held at Rottach-Egern, Germany, in 1980; at Rauischholzhausen, Germany, in 1983; and at Nairobi, Kenya, in 1986. The proceedings from each are available from the German Agency for Technical Cooperation, Deutsche Gesellschaft für Technische Zusammenarbeit (GTZ), Eschborn, Germany. The citations are as follows:

Schmutterer, H., K.R.S. Ascher, and H. Rembold, eds. 1981. *Natural Pesticides from the Neem Tree* (Azadirachta indica *A. Juss.).*

Schmutterer, H. and K.R.S. Ascher, eds. 1984. *Natural Pesticides from the Neem Tree* (Azadirachta indica *A. Juss.) and Other Tropical Plants.*

Schmutterer, H. and K.R.S. Ascher, eds. 1987. *Natural Pesticides from the Neem Tree* (Azadirachta indica *A. Juss.) and Other Tropical Plants.*

Two neem conferences have been held recently in the United States:

Locke, J.C. and R.H. Lawson, eds. 1990. *Neem's Potential in Pest Management Programs, Proceedings of the USDA Neem Workshop.* United States Department of Agriculture, Agricultural Research Service, ARS-86, 136 pp. Copies are available from James Locke, Plant Pathologist, Florist and Nursery Crops Laboratory, USDA/ARS, 10300 Baltimore Avenue, Beltsville, Maryland 20705-2350, USA.

Ahmed, S., ed. In press. Neem (*Azadirachta indica*) for pest control and rural development in Asia and the Pacific. Special session on neem from the 17th Pacific Science Congress, May 27 to June 2, 1991.

NEWSLETTER

Since 1984, a team of dedicated Indian scientists[1] have published the *Neem Newsletter*. This quarterly newsletter is supplied free of charge to those engaged in neem research and utilization. Copies are available from the Division of Agricultural Chemicals, Indian Agricultural Research Institute, New Delhi 110 012, India.

BIBLIOGRAPHY

At least one comprehensive bibliography of recent neem articles is available:

U.S. Department of Agriculture (USDA). 1989. *The Neem Tree: An Inhibitor of Insect Feeding and Growth, January 1982–April 1989*. NAL-BIBL. QB 89–89, Updates QB 86–27, July, 1989. Quick Bibliography Series. USDA, Beltsville, Maryland, USA.

THE VISION (Chapter 1)

A number of books and review articles describing neem and various aspects of its promise are available. These include the following:

Ahmed, S. 1985. *Utilizing Indigenous Plant Resources in Rural Development: Potential of the Neem Tree*. East-West Center, University of Hawaii, Honolulu, Hawaii.

Arnason, T.J., B.J.R. Philogène, and P. Morand, eds. 1988. *Insecticides of Plant Origin*. American Chemical Society (ACS) Symposium Series, Vol. 387. Symposium on Insecticides of Plant Origin at the Third Chemical Congress of North America, Toronto, Ontario, Canada, June 5–11, 1988. ACS, Washington, D.C.

Benge, M. 1986. *Neem: The Cornucopia Tree*. S&T/FENR Agro-Forestation Technical Series No. 5. Agency for International Development, Washington, D.C.

Cutler, H.G., ed. 1988. *Biologically Active Natural Products: Potential Use in Agriculture*. American Chemical Society (ACS) Symposium Series, Volume 380. Symposium on Biologically Active Natural Products, 194th meeting of the American Chemical Society, New Orleans, Louisiana, USA, August 30 to September 4, 1987. ACS, Washington, D.C. 483 pp.

Jacobson, M., ed. 1988. *1988 Focus on Phytochemical Pesticides: Volume 1, the Neem Tree*. CRC Press, Inc., Boca Raton, Florida, USA.

Ketkar, C.M. 1976. *Utilization of Neem (*Azadirachta indica *Juss.) and Its By-products*. Final technical report of Directorate of Non-Edible Oils and Soap Industry, Khadi and Village Industries Commission. Published by V. Lahshmikanthan, Irla Road, Vile Parle, Bombay 400 056.

Radwanski, S. 1977a. Neem tree 2: Uses and potential uses. *World Crops and Livestock* 29:111–113.

Radwanski, S. 1977b. Neem tree 3: Further uses and potential uses. *World Crops and Livestock* 29:167–168.

Radwanski, S.A. and G.E. Wickens. 1981. Vegetative fallows and potential value of neem tree in the tropics. *Economic Botany* 35(4):398–414.

Schmutterer, H. 1990. Properties and potential of natural pesticides from the neem tree *Azadirachta indica*. *Annual Review of Entomology* 35:271–297.

THE REALITY (Chapter 2)

Ali, B.H. 1987. The toxicity of *Azadirachta indica* leaves in goats and guinea pigs. *Veterinary and Human Toxicology* 29(1):16–19.

---

[1] These have included B.S. Parmar, editor; R.P. Singh, associate editor; K.N. Mehrotra; S.K. Bhatia; R. Prasad; and C. Devakumar.

Jotwani, M.C. and K.P. Srivastava. 1981. Neem: insecticide of the future. III. Chemistry, toxicology and future strategy. *Pesticides* 15(12):12.

Sadre, N.L., V.Y. Deshpande, K.N. Mendulkar, D.H. Nandal. 1984. Male antifertility activity of *Azadirachta indica* A. Juss (neem) in different species. Pages 473–482 in Schmutterer and Ascher, 1984 (see under above conference reports).

Sinniah, D. and G. Baskaran. 1981. Margosa oil poisoning as a cause of Reye's syndrome. *The Lancet* February 28, 1981:487–489.

Sinniah, D., G. Baskaran, L.M. Looi, and K.L. Leong. 1982. Reye-like syndrome due to margosa oil poisoning: report of a case with postmortem findings. *American Journal of Gastroenterology* 77(3):158–161.

Sinniah, D., P.H. Schwartz, R.A. Mitchell, and E.L. Arcinue. 1985. Investigation of an animal model of a Reye-like syndrome caused by margosa oil. *Pediatric Research* 19(21):1346–1355.

THE TREE (Chapter 3)

Bakshi, B.K. 1976. *Forest Pathology: Principles and Practice in Forestry*. Controller of Publications, Delhi.

Benge, 1986. (See above under chapter 1.)

CAB International Institute of Biological Control (CAB-IIBC). 1987. *Prospects for Biological Control of the Oriental Yellow Scale,* Aonidiella orientallis *(Newstead) as a Pest of Neem in Africa*. Ascot, Berkshire, UK.

Desai, S.G. et al. 1966. A new bacterial leaf spot and blight of *Azadirachta indica* A. Juss. *Indian Phytopathology* 19(3):322–3.

Roberts, H. 1965. *A Preliminary Check List of Pests and Diseases of Plantation Trees in Nigeria*. Federal Department of Forestry Research, Ibadan, Nigeria.

Sankaram, A.V.B., M.M. Murthy, K. Bhaskaraiah, M. Subramanyam, N. Sultana, H.C. Sharma, K. Leuschner, G. Ramaprasad, S. Sitaramaiah, C. Rukmini, and P.U. Rao. 1987. Chemistry, biological activity, and utilization of some promising neem extracts. Pages 127–148 in Schmutterer and Ascher, 1987 (see under above conference reports).

Schmutterer, H. 1990. Observations on pests of *Azadirachta indica* (neem tree) and of some *Melia* species. *Journal of Applied Entomology* 109:390–400.

Zech, W. 1984. Investigations on the occurrence of potassium and zinc deficiencies in plantations of *Gmelina arbores*, *Azadirachta indica* and *Anacardium occidentale* in semi-arid areas of West Africa. In *Potash Review, No. 1*. International Potash Institute, Berne, Switzerland.

WHAT'S IN A NEEM (Chapter 4)

Jacobson (1986b), Schmutterer (1984), and Jacobson et al. (1984) review the chemical work on seed oil extracts.

Shin-Foon Chiu. 1984. The active principles and insecticidal properties of some Chinese plants, with special reference to Meliaceae. Pages 255–262 in Schmutterer and Ascher, 1984 (see under above conference reports).

Cutler, 1988. (See above under chapter 1.)

Feuerhake, K.J. 1984. Effectiveness and selectivity of technical solvents for the extraction of neem seed components with insecticidal activity. Pages 103–114 in Schmutterer and Ascher, 1984 (see above under conference reports).

Jacobson, M. 1986a. Natural pesticides. Pages 144–148 in *Natural Resources: The 1986 Yearbook of Agriculture*. U.S. Government Printing Office, Washington, D.C.

Jacobson, M. 1986b. The neem tree: natural resistance par excellence. Pages 220–232 in *Natural Resistance of Plants to Insects*, ed. M.B. Green and P.A. Hedin. American Chemical Society (ACS) Symposium Series No. 296. ACS, Washington, D.C.

Jacobson, M. 1987. Neem research and cultivation in the western hemisphere. Pages 33–44 in Schmutterer and Ascher, 1987 (see above under conference reports).

Sanguanpong, U. and H. Schmutterer. In press. Laboratory trials on the effects of neem oil and neem-seed extracts against the two-spotted spider mite *Tetranychus urticae* Koch. *Zeitschrift für Pflanzenkrankheiten und Pflanzenschutz*.

Schmutterer, H. 1984. Neem research in the Federal Republic of Germany since the first international neem conference. Pages 21–30 in Schmutterer and Ascher, 1984 (see above under conference reports).

Stoll, G. 1986. *Natural Crop Protection, Based on Local Resources in the Tropics*. Josef Margraf, Publisher, Aichtal, Germany. 186 pp.

EFFECTS ON INSECTS (Chapter 5)

Adler, V.E. and E.C. Uebel. 1985. Effects of a formulation of neem extract on six species of cockroaches (Orthoptera: Blaberidae, Blattidae, and Blattellidae). *Phytoparasitica* 13(1):3–8.

Akou-Edi, D. 1984. Effects of neem seed powder and oil on *Tribolium confusum* and *Sitophilus zeamais*. Pages 445–451 in Schmutterer and Ascher, 1984 (see above under conference reports).

Arnason, J.T., B.J.R. Philogène, N. Donskov, M. Hudon, C. McDougall, G. Fortier, P. Morand, D. Gardner, J. Lambert, C. Morris, and C. Nozzolillo. 1985. Antifeedant and insecticidal properties of azadirachtin to the European corn borer, *Ostrinia nubilalis*. *Entomologia Experimentalis et Applicata* 38(1):29–34.

Champagne, D.E., M.B. Isman, and G.H.N. Towers. 1989. Insecticidal activity of phytochemicals and extracts of the Meliaceae. Pages 95–109 in Arnason et al., 1988 (see above under general reviews).

Devakumar, C., B.K. Goswami, and S.K. Mukerjee. 1985. Nematicidal principles from neem (*Azadirachta indica* A. Juss). I. Screening of neem kernel fractions against *Meloidogyne incognita* (Kofoid & White) Chitwood. *Indian Journal of Nematology* 15(1):121–124.

Lange, W. and K. Feuerhake. 1984. Increased activity of enriched neem seed extracts with synergist piperonyl butoxide under laboratory conditions. *Zeitschrift für Angewandte Entomologie* 98:368.

Locke and Lawson, 1990. (See above under conference reports.)

Mansour, F., K.R.S. Ascher, and N. Omari. 1987. Effects of neem (*Azadirachta indica*) seed kernel extracts from different solvents on the predacious mite *Phytoseiulus persimilis* and the phytophagous mite *Tetranychus cinnabarinus*. *Phytoparasitica* 15(2):125–130.

Plant Protection Directorate. n.d. *The Preservation of Beans (Cowpeas) with Neem Oil*. Technical Leaflet Plant Protection No. 3. Deutsche Gesellschaft für Technische Zusammenarbeit (GTZ) GmbH. B.P. 1263, Lomé-Cacaveli, Togo.

Plant Protection Directorate. n.d. *Treatment of Cabbage and Gboma Against Pests with Neem Seed Extract*. Technical Leaflet Plant Protection No. 2. Deutsche Gesellschaft für Technische Zusammenarbeit (GTZ) GmbH. B.P. 1263, Lomé-Cacaveli, Togo.

Rembold, H. 1989. Kairomones—chemical signals related to plant resistance against insect attack. Pages 352–264 in *New Crops for Food and Industry*, ed. G.E. Wickens, N. Haq, and P. Day. Chapman and Hall, London.

Romeo, J.T. and M.S.J. Simmonds. 1989. Nonprotein amino acids feeding deterrents from *Calliandra*. Pages 59–68 in Arnason et al., 1988 (see above under general reviews).

Saxena, R.C. 1989. Insecticides from neem. Pages 110–135 in Arnason et al., 1988 (see above under general reviews).

Saxena, R.C., P.B. Epino, Tu Cheng-Wen, and B.C. Puma. 1984. Neem, chinaberry and custard apple: antifeedant and insecticidal effects on leafhopper and planthopper pests of rice. Pages 403–412 in Schmutterer and Ascher, 1984 (see above under conference reports).

Schmutterer, 1990. (See above under chapter 1.)

Schmutterer, H. and T. Freres. 1990. Influence of neem-seed oil on metamorphosis, color and behavior of the desert locust, *Schistocerca gregaria* (Forsk.), and of the African migratory locust, *Locusta migratoria migratorioides* (R. & F.). *Zeitschrift für Pflanzenkrankheiten und Pflanzenschutz* 97(4):431–438.

Steffens, R.J. and H. Schmutterer. 1982. The effect of a crude methanolic neem (*Azadirachta indica*) seed kernel extract on metamorphosis and quality of adults of the Mediterranean fruit fly, *Ceratitis capitata* Wied. (Diptera, Tephutdae). *Zeitschrift für Angewandte Entomologie* 94:98–103.

von der Heyde, J., R.C. Saxena, and H. Schmutterer. 1984. Neem oil and neem extracts as potential insecticides for control of hemipterous rice pests. *Schrifteureiche GTZ* 1 61:377.

Warthen, Jr., J.D. 1989. Neem (*Azadirachta indica* A. Juss): organisms affected and reference list update. *Proceedings of the Entomological Society of Washington* 91(3):367–388.

Zehrer, W. 1984. The effect of the traditional preservatives used in Northern Togo and of neem oil for control of storage pests. Pages 453–460 in Schmutterer and Ascher, 1984 (see above under conference reports).

EFFECTS ON OTHER ORGANISMS (Chapter 6)

Devakumar et al., 1985. (See above under chapter 5.)

Grant, I.F. and H. Schmutterer. 1987. Effects of aqueous neem seed kernel extracts on ostracods (class Crustacea) development and population density in lowland rice fields. Pages 591–607 in Schmutterer and Ascher, 1987 (see above under conference reports).

Hoelmer, K.A., L.S. Osborne, and R.K. Yokomi. 1990. Effects of neem extracts on beneficial insects in greenhouse culture. Pages 100–102 in Locke and Lawson, 1990 (see above under conference reports).

Indian Leaf Tobacco Co. Ltd. *Tobacco Mosaic—Its Control with Neem Leaf Decoction*. Pamphlet No. 24 RJY 6'73 3,000 Indian Leaf Tobacco Co. Ltd.

Isman, M.B., D.T. Lowery, and O. Koul. In press. Laboratory and field evaluations of neem for control of aphid and lepidopteran pests. In *Resources for Sustainable Agriculture: The Use of Neem and Other Plant Materials for Pest Control and Rural Development*. Proceedings of the Symposium, XVII Pacific Science Congress, Honolulu, Hawaii, May 26 to June 2, 1991.

Ketkar, C.M. and M.S. Ketkar. 1984. Potential of neem oil and neem cake production in India. *Proceedings of Research Planning Workshop Botanical Pest Control Project, August 6–10, 1984, Los Baños, Philippines*. International Rice Research Institute (IRRI), Los Baños, Philippines. 30 pp.

Locke, J.C. 1990. Activity of extracted neem seed oil against fungal plant pathogens. Pages 132–136 in Locke and Lawson, 1990 (see above under conference reports).

Mansour, F., K.R.S. Ascher, and N. Omari. 1987. Effect of neem seed kernel extracts from different solvents on the predacious mite *Phytoseiulus persimilis* and the phytophagous mite *Tetranychus cinnabarinus* as well as on the predatory spider *Chiracanthium mildei*. Pages 577–587 in Schmutterer and Ascher, 1987 (see above under conference reports.)

Mariappan, V. and R.C. Saxena. 1984. Custard apple oil, neem oil and their mixtures: effect on survival of *Nephotettix virescens* and on rice tungro virus transmission. Pages 413–429 in Schmutterer et al., 1981 (see above under conference reports).

Muley, E.V. 1978. Biological and chemical control of the snail vector *Melania scabra* (Gastropoda: Prosobrachia). *Bulletin of the Zoological Survey of India* 1:1–5.

Rössner, J. and C.P.W. Zebitz. 1987a. Effect of neem products on nematodes and growth of tomato (*Lycopersicon esculentum*) plants. Pages 611–621 in Schmutterer and Ascher, 1987 (see above under conference reports).

Rössner, J. and C.P.W. Zebitz. 1987b. Effect of soil treatment with neem products on earthworms (Lumbricidae). Pages 627–632 in Schmutterer and Ascher, 1987 (see above under conference reports).

Saxena, R.C. 1987. Neem seed derivatives for management of rice insect pests: a review of recent studies. Pages 81–93 in Schmutterer and Ascher, 1987 (see above under conference reports).

Saxena, R.C., N.J. Liquido, and H.D. Justo. 1981. Neem seed oil, a potential antifeedant for the control of the rice brown planthopper *Nilaparvata lugens*. Pages 171–188 in Schmutterer et al., 1981 (see above under conference reports).

Saxena, R.C., H.D. Justo, Jr., and P.B. Epino. 1984. Evaluation and utilization of neem cake against the rice brown planthopper, *Nilaparvata lugens* (Homoptera: Delphacidae). *Journal of Economic Entomology* 77:502–507.

Saxena, R.C., Z.R. Khan, and N.B. Bajet. 1987. Reduction of tungro virus transmission by *Nephotettix virescens* (Homoptera: Cicadellidea) in neem cake-treated rice seedlings. *Journal of Economic Entomology* 80:1079–1082.

Schmutterer, H. and H. Holst. 1987. On the effect of enriched and formulated neem seed kernel extract AZT-VR-K on the honeybee *Apis mellifera*. *Zeitschrift für Angewandte Entomologie* 103:208–213.

Singh, R. 1971. Inactivation of potato virus X by plant extracts. *Phytopathologia Mediterranea* 10:211–213.

Singh, U.P., H.B. Singh, and R.B. Singh. 1980. The fungicidal effect of neem (*Azadirachta indica*) extracts on some soil-borne pathogens of gram (*Cicer arietinum*). *Mycologia* 72:1077–1093.

Singh, S.P., V. Pant, A.M. Khan, and S.K. Saxena. 1985. Changes in the phenolic contents, related rhizosphere mycoflora, and nematode population in tomato inoculated with *Meloidogyne incognita* as a result of soil amendment with organic matter. *Indian Journal of Nematology* 15(2):197–201.

Tripathi, R.K.R. and R.N. Tripathi. 1982. Reduction in bean common mosaic virus (BCMV) infectivity vis-a-vis crude leaf extract of some higher plants. *Experientia* 38:349.

MEDICINALS (Chapter 7)

Badam, L., R.P. Deolankar, M.M. Kulkarni, B.A. Nagsampgi, and U.V. Wagh. 1987. In vitro antimalarial activity of neem (*Azadirachta indica* A. Juss) leaf and seed extracts. *Indian Journal of Malariology* 24:111–117.

Elvin-Lewis, M. 1980. Plants used for teeth cleaning throughout the world. *Journal of Preventive Dentistry* 6:61–70.

Gandhi, M., R. Lal, A. Sankaranarayanan, C.K. Banerjee, and P.L. Sharma. 1988. Acute toxicity study of the oil from *Azadirachta indica* seed (neem oil). *Journal of Ethnopharmacology* 23:39–51.

Ganesalingham, V.K. 1987. Use of the neem plant in Sri Lanka at the farmer's level. Pages 95–100 in Schmutterer and Ascher, 1987 (see above under conference reports).

Garcia, E.S., P. Azambuja, H. Forester, and H. Rembold. 1984. *Zeitschrift für Naturforschung* 39C:1155.

Gill, J.S. 1972. *Studies on Insect Feeding Deterrents with Special Reference to the Fruit Extracts of the Neem Tree,* Azadirachta indica *A. Juss.* Ph.D. thesis, University of London. 260 pp.

Henkes, R. 1986. The neem tree: a farmer's friend. *The Furrow* October 1986:16.

Khalid, S.A., A. Farouk, T.G. Geary, and J.B. Jensen. 1986. Potential antimalarial candidates from African plants: an in-vitro approach using *Plasmodium falciparum*. *Journal of Ethnopharmacology* 15:201–209.

Khalid, S.A., H. Duddeck, and M. Gonzalez-Sierra. 1989. Isolation and characterization of an antimalarial agent of the neem tree *Azadirachta indica*. *Journal of Natural Products* 52(2):922–926.

Khan, M. and S.W. Wassilew. 1987. The effect of raw material from the neem tree, neem oil, and neem extracts on fungi pathogenic to humans. Pages 645–650 in Schmutterer and Ascher, 1987 (see above under conference reports).

Koul, O., M.B. Isman, and C.M. Ketkar. 1990. Properties and uses of neem, *Azadirachta indica*. *Canadian Journal of Botany* 68:1–11.

Nath, K., D.K. Agrawal, Q.Z. Hasan, S.J. Daniel, and V.R.B. Sastry. 1989. Water-washed neem (*Azadirachta indica*) seed kernel cake in the feeding of milch cows. *Animal Production* 48:497–502.

Patel, R.P. and B.M. Trivedi. 1962. The in-vitro antibacterial activity of some medicinal oils. *Indian Journal of Medical Research* 50:218–222.

Rae, A. and M.S. Sethi. 1972. Screening of some plants for their activity against vaccinia and fowlpox viruses. *Indian Journal of Animal Science* 42:1066–1070.

Rao, A.R., S.S.U. Kumar, T.B. Paramasivam, S. Kamalakshi, A.R. Parashuraman, and M. Shantha. 1969. Study of antiviral activity of tender leaves of margosa tree (*Melia azadirachta*) on vaccinia and variola virus: a preliminary report. *Indian Journal of Medical Research* 57:495–502.

Sadre et al. 1984. (See above under chapter 2.)

Schneider, B.H. 1986. The effect of neem leaf extracts on *Epilachna varivestis* and *Staphylococcus aureus*. Page 73 in *Abstracts of the 3rd International Neem Conference*, Nairobi, Kenya.

Sinha, K.C., S.S. Riar, R.S. Tiwary, A.K. Dhawan, J. Bardhan, P. Thomas, A.K. Kain, and R.K. Jain. 1984. Neem oil as a vaginal contraceptive. *Indian Journal of Medical Research* 79:131–136.

Sinniah and Baskaran, 1981. (See above under chapter 2.)

INDUSTRIAL PRODUCTS (Chapter 8)

Anderson, D.M.W. and A. Hendrie. 1971. The proteinaceous gum from *Azadirachta indica* A. Juss. *Carbohydrate Research* 20:259–268.

Anderson, D.M.W., A. Hendrie, and A.C. Munro. 1972. The amino acid composition of some plant gums. *Phytochemistry* 11:579–580.

Bringi, N.V. and M.S. Thakur. 1987. Neem (*Azadirachta indica* Juss) seed oil. Pages 118–142 in *Non-Traditional Oilseeds and Oils in India*, ed. N.V. Bringi. Oxford & IBH Publishing Co. Pvt. Ltd., New Delhi.

Mitra, C.R. 1963. *Neem*. Indian Central Oil Seeds Committee, Hyderabad, India.

Patrao, M.R. 1985. Rare neem tree with sweet leaves. *Neem Newsletter* 2(3):34.

Radwanski, S.A. and G.E. Wickens. 1981. Vegetative fallows and potential value of the neem tree (*Azadirachta indica*) in the tropics. *Economic Botany* 35(4):398–414.

Redknap, R.S. 1981. The use of crushed neem berries in the control of some insect pests in Gambia. Pages 205–214 in Schmutterer et al., 1981 (see above under conference reports).

Sarkar, M.S. and P.C. Datta. 1986. Biosynthesis of beta sitosterol in-vitro culture of *Azadirachta indica* cotyledon tissues. *Indian Drugs* 23(8).

REFORESTATION (Chapter 9)

Ahmed, S., S. Bamofleh, and M. Munshi. 1989. Cultivation of neem (*Azadirachta indica*, Meliaceae) in Saudi Arabia. *Economic Botany* 43:35–38.

Radwanski, S.A. 1969. Improvements of red acid sands by neem tree (*Azadirachta indica*) in Sokoto, North-West State Nigeria. *Journal of Applied Ecology* 6:507–511.

NEXT STEPS (Chapter 10)

Ahmed, S. 1990. *Symposium on Natural Resources for a Sustainable Agriculture, proceedings of a meeting, New Delhi, February 6–10, 1990*, ed. R.P. Singh. Indian Society of Agronomy, New Delhi.

Ahmed, S., M. Grainge, J.W. Hylin, W.C. Mitchel, and J.A. Litsinger. 1984. Some promising plant species for use as pest control agents under traditional farming systems. Pages 565–580 in Schmutterer and Ascher, 1984 (see above under conference reports).

Larson, R. 1987. Development of Margosan-O®, a pesticide from neem seed. Pages 243–250 in Schmutterer and Ascher, 1987 (see above under conference reports).

Michel-Kim, H. and A. Brandt. 1981. The cultivation of neem and processing it in a small village plant. Pages 279–90 in Schmutterer et al., 1981 (see above under conference reports).

Schmutterer et al., 1981. (See above under conference reports.)

Schmutterer and Ascher, 1984. (See above under conference reports.)

Schmutterer and Ascher, 1987. (See above under conference reports.)

APPENDIX D

RESEARCH CONTACTS

ASIA

Bangladesh

B.N. Islam, Department of Entomology, Faculty of Agriculture, Bangladesh Agricultural University, Mymensingh

India

S.K. Bhatia, Division of Agricultural Chemicals, Indian Agricultural Research Institute, New Delhi 110 012

N.V. Bringi, Chemical Sciences Group, Hindustan Lever Research Centre, Express Building, 2 Bahadur Shah Zafar Marg, New Delhi 110 002

Calcutta Chemical Co. Ltd., 35 Panditia Road, Calcutta-29

Coimbatore Post-Graduate Centre, c/o University of Madras, Chipauk, Triplicane P.O., Madras, Tamil Nadu, 600 005 (seed collection)

S.C. Das, Plant Protection Department, Tocklai Experiment Station, Jorhat, Assam

C. Devakumar, Division of Agricultural Chemicals, Indian Agricultural Research Institute, New Delhi 110 012

Excelsior Enterprises, B-12, A.M. Jaipuria Road Cantt., Kanpur 208 004, Uttar Pradesh (commercial contraceptive)

B. Gope, Plant Protection Department, Tocklai Experiment Station, Jorhat, Assam

S. Jayaraj, Centre for Plant Protection Studies, Tamil Nadu Agricultural University, Coimbatore 641 003

B.G. Joshi, Central Tobacco Research Institute, Rajahmundry, Andhra Pradesh 533 105

A. Abdul Kareem, Department of Entomology, CIP R&D Farm, Padappai, Chigleput (Dist.), PIN 601 301

Charu Kaushic, National Institute of Immunology, New Mehrauli Road, New Delhi 110 067

C.M. Ketkar, Neem Mission, 471, Shaniwar Peth, Pune 411 030, Maharashtra

T.N. Khoshoo, Tata Energy Research Institute (teri), 7 Jor Bagh, New Delhi 110 003

K.N. Mehrotra, Division of Agricultural Chemicals, Indian Agricultural Research Institute, New Delhi 110 012

S. Mukherjee, Plant Protection Department, Tocklai Experiment Station, Jorhat, Assam

B.S. Parmar, Division of Agricultural Chemicals, Indian Agricultural Research Institute, New Delhi 110 012

Rajendra Prasad, Division of Agricultural Chemicals, Indian Agricultural Research Institute, New Delhi 110 012

S.S. Riar, Defence Institute of Physiology and Allied Sciences, Delhi Cantonment 110 010

A.V.B. Sankaram, Division of Organic Chemistry, Regional Research Laboratory, Hyderabad 500 007

Rita Shah, Toxicology Research Foundation, KIM, 2057 Sadashiv, Vijayanagar Colony, Pune 411 030

R.N. Sharma, National Chemical Laboratory, Pune 411 008, Maharashtra

R.P. Singh, Division of Entomology, Indian Agricultural Research Institute, New Delhi 110 012

S.P. Singh, Department of Botany, Aligarh Muslim University, Aligarh 202 001, Uttar Pradesh

K.C. Sinha, Defence Institute of Physiology and Allied Sciences, Delhi Cantonment 110 010

G.P. Talwar, National Institute of Immunology, JNU Complex, Shahid Jeet Singh Marg, New Delhi 110 067

Shaktin Upadhyay, National Institute of Immunology, JNU Complex, Shahid Jeet Singh Marg, New Delhi 110 067

R.R. Verma, Government Valley Fruit Research Station, Srinagar, Garhwal

Indonesia

Mumu Sutisna, Interuniversity Center on Life Sciences, Department of Biology, Jl. Ganesha 10 Bandung, 40132

Malaysia

D. Sinniah, Department of Pediatrics and Central Animal House Faculty of Medicine, University of Malaya, Lembah Pantai, 59100 Kuala Lumpur

Pakistan

S.N.H. Naqvi, Department of Zoology, University of Karachi, Karachi 32

People's Republic of China

Shin-Foon Chiu, Laboratory of Insect Toxicology, Department of Plant Protection, South China Agricultural University, Guangzhou

Philippines

International Rice Research Institute (IRRI), P.O. Box 933, Manila

G.S. Karganilla, Bureau of Plant Industry, Crop Protection Division, 692 San Andres St., Malate, Manila

Pekka K. Ketola, Consultant, Manila Seedling Bank Foundation, Inc., Quezon Boulevard, corner EDSA, Quezon City

Sri Lanka

V.K. Ganesalingam, Department of Zoology, University of Jaffna, Thirunelvely, Jaffna

Thailand

K. Sombatsiri, Department of Entomology, Kasetsart University, Bangkok 10 900

PACIFIC

Australia

A. Bösselmann, Department of Botany, University of Queensland, 4072 Queensland
Neemoil Australia, Pty. Ltd., 88 Habib Drive, Lismore, 2480 New South Wales
Martin J. Rice, Department of Entomology, University of Queensland, St. Lucia, Brisbane, 4067 Queensland

New Zealand

Gordon Grandison, DSIR Plant Protection, Mt. Albert Research Centre, 120 Mt. Albert Road, Private Bag, Auckland

AFRICA

Ghana

Edward S. Ayensu, Pan-African Union for Science and Technology, Ghana, P.O. Box 16525 AIRPORT, Accra
B. Tanzubil, Nyankpala Agricultural Experiment Station, P.O. Box 483, Tanale, Accra

Kenya

J. Achula, International Centre of Insect Physiology and Ecology (ICIPE), P.O. Box 30772, Nairobi
Kofi Atsesekapo, Kenya Medical Research Institute (KEMRI), P.O. Box 54840, Nairobi (malaria)
A. Chapya, International Centre of Insect Physiology and Ecology (ICIPE), P.O. Box 30772, Nairobi
I. Griesbach, GAT/GTZ, P.O. Box 14272, Nairobi
A. Hassanali, International Centre of Insect Physiology and Ecology (ICIPE), P.O. Box 30772, Nairobi
T.T.O. Isaac, International Centre of Insect Physiology and Ecology (ICIPE), P.O. Box 30772, Nairobi
H. Jästedt, Forestry Project Mubuga, GTZ-Project Administration Service (PAS), P.O. Box 41607, Nairobi
J.N. Kaman, Pyrethrum Board of Kenya, P.O. Box 420, Nakuru
Margaret W. Kinuthia, Coffee Research Foundation, Coffee Research Station, P.O. Box 4, Ruiru
J.N. Kuria, Pyrethrum Board of Kenya, P.O. Box 420, Nakuru
W. Lwande, International Centre of Insect Physiology and Ecology (ICIPE), P.O. Box 30772, Nairobi
M.J. Mutinga, International Centre of Insect Physiology and Ecology (ICIPE), P.O. Box 30772, Nairobi
R.W. Mwangi, Department of Zoology, University of Nairobi, P.O. Box 30197, Nairobi
E. Nyandat, International Centre of Insect Physiology and Ecology (ICIPE), P.O. Box 30772, Nairobi
Rosemary Okott, International Centre of Insect Physiology and Ecology (ICIPE), P.O. Box 30772, Nairobi

D.A. Otieno, International Centre of Insect Physiology and Ecology (ICIPE), P.O. Box 30772, Nairobi
Ramesh C. Saxena, IPMB, Rice Project, International Centre of Insect Physiology and Ecology (ICIPE), P.O. Box 30772, Nairobi
J.O.I. Tondiko, International Centre of Insect Physiology and Ecology, P.O. Box 30772, Nairobi
B. Torto, International Centre of Insect Physiology and Ecology (ICIPE), P.O. Box 30772, Nairobi
F.M.E. Wanjala, Coffee Research Foundation, P.O. Box 4, Ruiru

Mauritius

I. Fagoonee, School of Agriculture, University of Mauritius, Réduit

Nigeria

D.E.U. Ekong, University of Port Harcourt, East-West Road, Choba, P.M.B. 5323, Port Harcourt
H. Herren, International Institute of Tropical Agriculture (IITA), Ibadan
M.F. Ivbijaro, Department of Agricultural Biology, University of Ibadan, Ibadan
J.I. Olaifa, Department of Plant Science, University of Ife, Ile-Ife

Rwanda

Edouard Niziyimana, OPROVIA-GRENARWA II—Recherches, National Food Quality Laboratory, Office National pour le Developpement et la Commercialisation des Produits Vivriers et des Production Animales, Kigali
Aussumani Serugendo, OPROVIA-GRENARWA II—Recherches, National Food Quality Laboratory, Office National pour le Developpement et la Commercialisation des Produits Vivriers et des Production Animales, Kigali

Senegal

Saliou Diangar, Agronome/Programme Mil, Centre National de Recherches Agronomiques (CNRA), Institut Sénégalais de Recherches Agricoles (ISRA), Boîte Postal 53, Bambey

Tanzania

A.K. Karel, Department of Crop Science, Sokoine University of Agriculture, P.O. Box 3042, Morogoro
Hildegard Keil, Tropical Pesticides Research Institute, P.O. Box 3024, Arusha
D.H. Matemu, Tropical Pesticides Research Institute, P.O. Box 3024, Arusha
J.A. Saidi, Tropical Pesticides Research Institute, P.O. Box 3024, Arusha

Togo

S. Adhikary, Sce. Protection des Végétaux, Projet Allemand, B.P. 1263, Lomé-Cacaveli
Plant Protection Directorate, B.P. 1263, Lomé-Cacaveli
W. Zeher, Sce. Protection des Végétaux, Projet Allemand, B.P. 1263, Lomé-Cacaveli

EUROPE

Belgium

F. Colin, Biochem Products S.A., 479 Avenue Louise, 1050 Brussels
A.Y. Le Vernoy, Biochem Products S.A., 479 Avenue Louise, 1050 Brussels

France

Ronald Bellefontaine, Centre Technique Forestier Tropical, Département du CIRAD, 45 bis, avenue de la Belle Gabrielle, 94736 Nogent-sur-Marne Cedex
Centre Technique Forestier Tropical (CTFT), 45 bis, avenue de la Belle Gabrielle, 94736 Nogent-sur-Marne Cedex
Jean E. Gorse, 13, avenue Maréchal, Franchet d'Esperey, 75016 Paris, France
Christine Holding, Usclas de Bosc, Lodere 34700
S.A. Radwanski, 4, rue Georges Bergers, 75017 Paris 17
Y. Roederer, Centre Technique Forestier Tropical, Département du CIRAD, 45 bis, avenue de la Belle Gabrielle, 94736 Nogent-sur-Marne Cedex
D. Le Rumeur, Roussel UCLAF, 163 avenue Gambetta, Paris

Germany

H. Adolphi, BASF Aktiengesellschaft, Landwirtschaftliche Versuchsstation, Postfach 220, 6703 Limburgerhof
H.-J. Bidmon, Institut für Physiologische Chemie I, Philipps-Universität, Deutschhausstrasse 1—2, 3550 Marburg
D. Biedenkopf, Institut für Phytopathologie und Angewandte Zoologie, Justus-Liebig-Universität, Ludwigstrasse 23, 6300 Giessen
M. Bokel, Institut für Chemie, Universität Hohenheim, Garbenstrasse 30, 7000 Stuttgart 70
R. Bürstinghaus, Hauptlaboratorium, BASF AG, 6700 Ludwigshafen
Ch. Czoppelt, Max-Planck-Institut für Biochemie, 8033 Martinsried b. München
A. Dorn, Institut für Zoologie, Fachbereich Biologie, Johannes-Gutenberg Universität, Postfach 3980, 6500 Mainz
M. Dreyer, Institut für Phytopathologie und Angewandte Zoologie, Justus-Liebig-Universität, Ludwigstrasse 23, 6300 Giessen
Helmut Duddeck, Rubr-Universität Bochum, Fakultät für Chemie, Postfach 10 21 48, D-4630 Bochum 1
K. Ermel, Institut für Pflanzenphysiologie, Justus-Liebig-Universität, Heinrich-Buff-Ring 54—62, 6300 Giessen
A. Jäger, Institute of Agrochemical Research, Biological Department, Schering AG, Gollanczstrasse 71-99, 1000 Berlin 28
M. Jung, Gesellschaft für Technische Zusammenarbeit (GTZ) GmbH, Postfach 5180, 6236 Eschborn 1
C. Jürgens, Gesellschaft für Technische Zusammenarbeit (GTZ), Dag-Hammarskjold-Weg 1, 6236 Eschborn 1
K. Kirsch, Institut für Phytopathologie und Angewandte Zoologie, Justus-Liebig-Universität, Ludwigstrasse 23, 6300 Giessen
Inge Kleinschmit, Institut für Phytopathologie und Angewandte Zoologie, Justus-Liebig-Universität, Ludwigstrasse 23, 6300 Giessen
W. Kraus, Institut für Chemie, Universität Hohenheim, Garbenstrasse 30, 7000 Stuttgart 70
W. Lange, Institut für Phytopathologie und Angewandte Zoologie, Justus-Liebig-Universität, Ludwigstrasse 23, 6300 Giessen
M. Michel-Kim, Eco-Region GmbH, Consulting Firm, Bamberger Strasse 41, 1000 Berlin 30

Heinz Rembold, Max-Planck-Institut für Biochemie, D-8033 Martinsried bei München
Marlies Schauer, Institut für Phytopathologie und Angewandte Zoologie, Justus-Liebig-Universität, Ludwigstrasse 23, 6300 Giessen
U. Schlüter, Institut für Allgemeine und Spezielle Zoologie, Justus-Liebig-Universität, Stephanstrasse 24, 6300 Giessen
G.H. Schmidt, Lehrgebiet Zoologie-Entomologie, Universität Hannover, Herrenhäuser Strasse 2, 3000 Hannover 21
Heinrich Schmutterer, Institut für Phytopathologie und Angewandte Zoologie, Justus-Liebig-Universität, Ludwigstrasse 23, 6300 Giessen
F.A. Schulz, Institut für Phytopathologie, Christian-Albrechts-Universität, Olshausenstrasse 40, 2300 Kiel
Martina Schwinger, Institut für Chemie, Universität Hohenheim, Garbenstrasse 30, 7000 Stuttgart 70
K.P. Sieber, Max-Planck-Institut für Biochemie, 8033 Martinsried b. München
G.K. Sharma, Max-Planck-Institut für Biochemie, 8033 Martinsried b. München
R. Steets, Verkauf Landwirtschaft-Beratung, Höchst AG, Postfach 800320, 6230 Frankfurt A.M. 80
M. Völlinger, Institut für Phytopathologie und Angewandte Zoologie, Justus-Liebig-Universität, Ludwigstrasse 23, 6300 Giessen
H. Wilps, Institut für Biologie I (Zoologie), Albert-Ludwigs-Universität, Albertstrasse 21 a, 7800 Freiburg
R. Wohlgemuth, Institut für Vorratsschutz, Biologische Bundesanstalt, Königin-Luise-Strasse 19, 1000 Berlin 33
C.P.W. Zebitz, Institut für Pflanzenkrankheiten und Pflanzenschutz, Universität Hannover, 3000 Hannover 21
G. Zoebelein, Sparte Pflanzenschutz-Anwemdungstechnik, Biologische Forschung, Bayer AG, Leverkusen, Bayerwerk

Netherlands

L.M. Schoonhover, Agricultural University, Salverdaplein 11, POB 9101, 6700 HB, Wageningen (insect behavior, receptors)

Switzerland

R. Lamb, IRRI/Swiss Development Cooperation, Beethovenstrasse 11, 3073 Gümlingen
H.R. Waespe, Ciba-Geigy AG, CH-4002 Basel

United Kingdom

Robert W. Fishwick, The Red House, North Bovey, Devon TQ13 8RA
Ian F. Grant, Natural Resources Institute, Central Avenue, Chatham Maritime, Kent ME4 4TB
Philip S. Jones, Roche Products, Ltd., P.O. Box 8, Welwyn Garden City, Hertfordshire AL3 3AY
Steven V. Ley, Department of Chemistry, Imperial College of Science and Technology, South Kensington, London SW7 2AZ
E. David Morgan, Department of Chemistry, University of Keele, Staffordshire ST5 5BG
R.S. Redknap, 44, Molesworth House, Royal Road, Kensington, London SE 17
Dinos Santafianos, Department of Chemistry, Imperial College of Science and Technology, South Kensington, London SW7 2AZ
Gerald E. Wickens, 50 Uxbridge Road, Hampton Hill, Middlesex TW12 3AD

MIDDLE EAST

Israel

K.R.S. Ascher, Institute of Plant Protection, Department of Toxicology, The Volcani Center, Agricultural Research Organization, P.O. Box 6, Bet Dagan 50-250

F. Mansour, Agricultural Research Organization, Regional Experiment Station, Newe Ya'ar, P.O. Haifa

J. Meisner, Institute of Plant Protection, The Volcani Center, Agricultural Research Organization, P.O. Box 6, Bet Dagan 50-250

The Sudan

Sami A. Khalid, Department of Pharmacognosy, University of Khartoum, P.O. Box 1996, Khartoum

Tigani El Mohd, El Amin, Agricultural Research Corporation, P.O. Box 126, Wad Medani

A.S. Siddig, Agricultural Research Corporation, Shambat Research Station, P.O. Box 30, Khartoum North

CENTRAL AND SOUTH AMERICA

Argentina

Manuel Gonzales-Sierra, IQUIOS, Universidad Nacional de Rosario, Suipacha 531, 2000 Rosario

Brazil

Eloi de Souza Garcia, Department of Entomology, Department of Biochemistry and Molecular Biology, Fundaçào Oswald Cruz, Av. Brasil 4365, CP 926, 21040 Rio de Janeiro

Costa Rica

Centro Agronómico Tropical de Investigación y Enseñanza (CATIE), Turrialba

R.A. Vargas, Ministry of Agriculture, Dirección de Sanidad Vegetal-Ap., 10094-1000 San José

Ecuador

C. Klein-Koch, Asistencia Tecnica Alemana, Casilla 11.368, C.C.N.U., Quito

Haiti

Peter Welle, Coordinator, Farmers' Resources Management Project, CARE-Haiti, B.P. 773, Port-au-Prince

Nicaragua

J. Mercado, Projecto Fortalacimiento del Servicio Fitosanitario, Apartado 489, Managua

NORTH AMERICA

Canada

John T. Arnason, Department of Biology, 550 Cumberland Street, University of Ottawa, Ottawa, Ontario K1N 6N5

Murray B. Isman, Department of Plant Science, University of British Columbia, Vancouver, British Columbia V6T 2A2

Jeff Stewart, Agriculture Canada Research Station, Charlottetown, Prince Edward Island C1A 7M8

United States

The Neem Association, 1511 Oneco Avenue, Winter Park, Florida. (Publishes a bimonthly newsletter, *Neem News*.)

AgriDyne Technologies, Inc., AgriDyne Technologies, Inc., 2401 South Foothill Drive, Salt Lake City, Utah 84109 (commercial products)

Grace/Sierra Horticultural Products, 570 Grant Way, Fogelsville, Pennsylvania 18051 (commercial products)

Ringer, Inc., Valley View Road, Eden Prairie, Minnesota 55344-3585 (commercial products)

Saleem Ahmed, East-West Center, 1777 East-West Road, Honolulu, Hawaii 96848

Agricultural Research Service, 13601 Old Cutler Road, U.S. Department of Agriculture, Miami, Florida 33158

David Akey, Agricultural Research Service, U.S. Department of Agriculture, 413 South East Broadway, Phoenix, Arizona 85040

James R. Baker, Department of Entomology, North Carolina State University, Raleigh, North Carolina 27695-7613

Mohammad Bari, Artichoke Research Association, Agriculture Research Station, 1636 East Alisal Street, Salinas, California 93905

Joseph W. Begley, Yoder Brothers, Inc., PO Box 68, Alva, Florida 33920

Michael D. Benge, Office of Environment and Natural Resources, U.S. Agency for International Development, Washington, DC 20523 (ecology and management)

Jo-Ann Bentz, Florist & Nursery Crops Laboratory, U.S. Department of Agriculture, Building 470 Room 10, Beltsville Agricultural Research Center (BARC)-East, Beltsville, Maryland 20705

Deepak Bhatnagar, Food and Feed Safety Research, Southern Regional Research Center, U.S. Department of Agriculture, PO Box 19687, New Orleans, Louisiana 70179

Baruch Blumberg, Associate Director for Clinical Research, Division of Population Oncology, Fox Chase Cancer Center, 7701 Burholme Avenue, Philadelphia, Pennsylvania 19111

John D. Briggs, Department of Entomology, The Ohio State University, 103 Botany and Zoology Building, 1735 Neil Avenue, Columbus, Ohio 43210-1220

Raymond Brush, American Association of Nurserymen, Inc., 1250 Eye Street, N.W., Suite 500, Washington, DC 20005

Richard Casagrande, Department of Plant Sciences, University of Rhode Island, Kingston, Rhode Island 02881-0804

John Conrick, 1511 Oneco Avenue, Winter Park, Florida (planting materials)

Whitney Cranshaw, Department of Entomology, Colorado State University, Fort Collins, Colorado 80523

Edward M. Croom, Jr., Research Institute of Pharmaceutical Sciences, Health Sciences Research Division, School of Pharmacy, The University of Mississippi, University, Mississippi 38677

Florence V. Dunkel, Entomology Research Laboratory, Montana State University, Bozeman, Montana 59717

Robert M. Faust, National Program Staff, U.S. Department of Agriculture, Building 005 Room 236, BARC-West, Beltsville, Maryland 20705

Dave Ferro, Department of Entomology, Fernald Hall, University of Massachusetts, Amherst, Massachusetts 01103

Glenn Fisher, Cordley Hall 2046, Oregon State University, Corvallis, Oregon 97331-2907

Harrison "Chick" Gardner, Milagro Farms, 25776 Honda Road, Madera, California 93638 (planting materials)

R.G. Gilbert, U.S. Department of Agriculture, PO Box 2890, Washington, DC 20013

Michael Grainge, East-West Center, 1777 East-West Road, Honolulu, Hawaii 96848

Robert Harwell, Pine Hill Farms, PO Box 7249, Winter Haven, Florida 33883 (planting materials)

Donald Heyneman, Department of Epidemiology and Biostatistics, School of Medicine, San Francisco, California 94143-0560 (schistosomiasis)

Kim Hoelmer, Horticultural Research Laboratory, U.S. Department of Agriculture, 2120 Camden Road, Orlando, Florida 23803

Bill Howard, U.S. Department of Agriculture, 3205 College Avenue, Fort Lauderdale, Florida 33314

Marlin Huffman, Plantation Botanicals, Inc., PO Box 128, Felda, Florida 33930 (planting materials)

Martin Jacobson, 1131 University Boulevard West, Apartment 616, Silver Spring, Maryland 20902

Subhash C. Juneja, Department of Obstetrics and Gynecology, College of Medicine, University of Florida, Gainesville, Florida 32610 (contraception)

Clifford Keil, Department of Entomology, University of Delaware, Newark, Delaware 19716-0701

Marc Ketchel, The Neem Company, Route 3 Box 32G, Alachua, Florida 32615 (planting materials)

Walter Knausenberger, Agriculture and Natural Resources Division, Office of Technical Resources, Bureau for Africa, Agency for International Development, Washington, DC 20523

T.L. Ladd Jr., USDA Japanese Bettle and Horticultural Insect Pests Research Laboratories, Ohio Agricultral Research and Development Center, Wooster, Ohio 44691

Hiram Larew, Policy Directorate, Office of Strategic Planning, Agency for International Development, Washington, DC 20523

Robert O. Larson, Vikwood Botanicals, Inc., PO Box 554, Sheboygan, Wisconsin 53082

Roger H. Lawson, Florist & Nursery Crops Laboratory, U.S. Department of Agriculture, Building 004 Room 208, BARC-West, Beltsville, Maryland 20705

David W. Lee, Department of Biological Sciences, College of Liberal Arts and Sciences, Florida International, The Public University at Miami, University Park, Miami, Florida

Gary Leibee, Central Florida Research and Education Center, 2700 East Celery Avenue, Sanford, Florida 32771

Daniel Leskovar, Agricultural Research and Extension Center, Texas A&M University, 1619 Garner Field Road, Uvalde, Texas 78801

Zev Lidert, Rohm and Haas Company, Spring House Research Center, Spring House, Pennsylvania 19477

Richard K. Lindquist, Department of Entomology, Ohio Agricultural R&D Center, Wooster, Ohio 44691

James C. Locke, Plant Pathology, Florist and Nursery Crops Laboratory, Agricultural Research Service, U.S. Department of Agriculture, 10300 Baltimore Avenue, Beltsville, Maryland 20705-2350

K. Maramorosch, Department of Entomology, Rutgers University, New Brunswick, New Jersey 08903

Ronald Mau, Department of Entomology Room 310, University of Hawaii, Honolulu, Hawaii 96822

Julius J. Menn, Plant Sciences Institute, U.S. Department of Agriculture, Beltsville, Maryland 20705

J. Allen Miller, Agricultural Research Service, U.S. Department of Agriculture, PO Box 232, Kerrville, Texas 78029 (hornfly)

Koji Nakanishi, Columbia University, Morningside Heights, New York, New York 10027 (chemistry of neem ingredients)

John W. Neal, Florist and Nursery Crops Laboratory, U.S. Department of Agriculture, Building 470 Room 8, BARC-East, Beltsville, Maryland 20705

David Nielsen, Department of Entomology, OARDC, The Ohio State University, Wooster, Ohio 44691-4096

Gregg Nuessly, Everglades Research and Education Center, PO Box 8003, Belle Glade, Florida 33430-8003

Ronald D. Oetting, Department of Entomology, Georgia Experiment Station, Experiment, Georgia 30212

L.S. Osborne, Central Florida Research and Education Center, 2807 Binion Road, Opopka, Florida 32703

Michael Parrella, Department of Entomology, University of California, Davis, California 95616-8584

John A. Parrotta, USDA Institute of Tropical Forestry, Call Box 25000, Rio Piedras, Puerto Rico 00928-2500 (ecology)

James F. Price, University of Florida, Agricultural Research and Development Center, 5007 60th Street East,, Bradenton, Florida 34203

Martin L. Price, Educational Concerns for Hunger Organization (ECHO), 17430 Durrance Road, North Ft. Myers, Florida 33903 (planting materials)

Edward B. Radcliffe, Department of Entomology, University of Minnesota, St. Paul, Minnesota 55108

R.E. Redfern, Livestock Insects Laboratory, Agricultural Environmental Quality Institute, U.S. Department of Agriculture, Beltsville, Maryland 20705

David Riley, Texas Agricultural Research Station, Texas A&M University, 2415 East Highway 83, Weslaco, Texas 78596

Fred Saleet, The Banana Tree, 715 Northampton Street, Easton, Pennsylvania 18042 (planting materials)

John P. Sanderson, Department of Entomology, Cornell University, Ithaca, New York 14853

Roland Seymour, Department of Plant Biology, The Ohio State University, 1735 Neil Avenue, Columbus, Ohio 43210-1220

Dave Shetlar, Entomology Extension, 1991 Kenny Road, Columbus, Ohio 43210

Eugene B. Shultz, Jr., c/o Department of Engineering and Applied Science, Campus Box 1106, Washington University, St. Louis, Missouri 63130

David Smitely, Department of Entomology, Michigan State University, East Lansing, Michigan 48824-1115

Shobha Sriharan, Division of Natural Sciences, Selma University, Selma, Alabama 36701

John D. Stark, Washington State University, 7612 Pioneer Way E., Puyallup, Washington 98371 (fruit fly and aphid control)

J. Rennie Stavely, Microbiology and Plant Pathology Laboratory, Agricultural Research Service, U.S. Department of Agriculture, Building 011A, BARC-West, Beltsville, Maryland 20705

Peter P. Strzok, Agency to Facilitate the Growth of Rural Organizations (AFGRO), PO Box 14926, University Station, Minneapolis, Minnesota 55414 (ecology and management)

Ward Tingey, Department of Entomology, Comstock Hall, Cornell University, Ithaca, New York 14853-0999

Nick C. Toscano, College of Natural Resources Programs, 305 College Building North, University of California, Riverside, California 92521

E.C. Uebel, Insect Chemical Ecology Laboratory, U.S. Department of Agriculture, Building 402E Room 108, BARC-East, Beltsville, Maryland 20705

Iroka J. Udeinya, Department of Immunology, Walter Reed Army Institute of Research, Washington, DC 20307

Diane E. Ullman, Department of Entomology, University of Hawaii, 3050 Maile Way, Honolulu, Hawaii 96822

David W. Unander, Division of Population Oncology, Fox Chase Cancer Center, 7701 Burholme Avenue, Philadelphia, Pennsylvania 19111 (medicinal uses)

P. Venkasawaran, Division of Population Oncology, Fox Chase Cancer Center, 7701 Burholme Avenue, Philadelphia, Pennsylvania 19111

Jim Walter, W.R. Grace Washington Research Center, 7379 Route 32, Columbia, Maryland 20861

David Warthen, Agricultural Research Service, U.S. Department of Agriculture, Building 007 Room 337, BARC-West, Beltsville, Maryland 20705

Ralph Webb, Insect Chemical Ecology Laboratory, U.S. Department of Agriculture, Building 402E Room 108, BARC-East, Beltsville, Maryland 20705

G.W. Zehnder, Eastern Shore Agricultural Research Station, Painter, Virginia 23420 (Colorado potato beetle control)

Eldon Zehr, Department of Plant Pathology and Physiology, Clemson University, Clemson, South Carolina 29634-0377 (nematodes)

APPENDIX E

Biographical Sketches of Panel Members

EUGENE B. SHULTZ, JR., *Chairman*, is professor of engineering and applied science at Washington University in St. Louis, Missouri, and director of the Bioresources Development Group at Washington University. He earned his B.S. degree in chemistry at Principia College and his M.S. and Ph.D. degrees in chemical engineering at the Illinois Institute of Technology. For 10 years, he was involved in research and development on solid, liquid, and gaseous fuels at the Institute of Gas Technology, Chicago, conducting laboratory and engineering-economic studies. He spent 15 years at Principia College, serving as chairman of the department of chemistry and as Kent H. Smith Professor of Chemistry. Since joining Washington University in 1979, his principal interests have included global environmental problems, Third World issues, and unconventional bioresources, mainly the development of renewable energy and appropriate technology, and the management of technological innovation in the Third World. In 1987, as a Fulbright researcher, he studied unconventional crops for food oils, high protein, fuel alcohol, and nontoxic botanic insect-control extracts at the University of Costa Rica. He has written numerous papers on dried roots for solid fuel and for fermentation to fuel alcohol and on unconventional seeds as new sources of edible and industrial oils. Currently, he serves as associate editor of *Economic Botany* for processing and utilization of economic plants. In 1991, he was elected both president-elect of the Association of Arid Lands Studies and program chair for its 1992 annual meeting. He also served on the program committee for the 1991 annual meeting of the Africa Studies Association.

DEEPAK BHATNAGAR is a geneticist with the U.S. Department of Agriculture's Agricultural Research Service at the Southern Regional Research Center in New Orleans, Louisiana. He received his B.Sc. from the University of Udaipur in India in 1972 and his M.Sc. and

Ph.D. from the Indian Agricultural Research Institute, New Delhi, in 1974 and 1977, respectively. From 1974 to 1977 he was a senior research fellow at the Indian Agricultural Research Institute, followed by work with the Department of Biophysics at the All India Institute of Medical Sciences, New Delhi, and the Department of Biology at Purdue University, West Lafayette, Indiana. From 1981 to 1985 he was a senior research associate with the Department of Biochemistry at Louisiana State University School of Medicine in New Orleans. From 1985 on he has worked on the USDA's project on bioregulatory control of aflatoxin biosynthesis. His major interests include the control of aflatoxin contamination of food and feed through an understanding of the molecular regulation of the biosynthesis of the toxin. He is a member of the American Chemical Society, the American Society for Biochemistry and Molecular Biology, the American Society of Plant Physiologists, and the American Society for Microbiology. He is a member of the editorial boards of *Applied and Environmental Microbiology* and *Mycopathologia*, and has edited several publications on mycotoxins and on improving food quality and safety.

MARTIN JACOBSON received his degree in chemistry from the City University of New York. From 1964 to 1972, Mr. Jacobson was an investigations leader with the U.S. Department of Agriculture's Entomological Research Division at Beltsville, Maryland; chief of the Biologically Active Natural Products Laboratory from 1973 to 1985; and research leader (plant investigations) with the Insect Chemical Ecology Laboratory until his retirement from federal service in 1986. He is currently an agricultural consultant in private practice in Silver Spring, Maryland. His awards include the Hillebrand Prize of the Chemical Society of Washington in 1971; USDA Certificates of Merit and cash awards for research in 1965, 1967, and 1968; the McGregory Lecture Award in Chemistry at Colgate University (Syracuse, New York); two bronze medals for excellence in research at the 3rd International Congress of Pesticide Chemistry, Helsinki, Finland, in 1974; USDA Director's Award on Natural Products Research in 1981; and an Inventor's Incentive Award for commercialization of a boll weevil deterrent in 1983. Mr. Jacobson is the author or coauthor of more than 300 scientific reports in numerous journals, the author of five books (*Insect Sex Attractants*, Wiley, 1965; *Insect Sex Pheromones*, Academic Press, 1972; *Insecticides from Plants: A Review of the Literature, 1941–1953*, USDA Handbook No. 154, 1958; *Insecticides from Plants: A Review of the Literature, 1954–1971*, USDA Handbook No. 461, 1975; *Glossary of Plant-Derived Insect Deterrents*, CRC Press, 1990); and editor of the books *Naturally Occurring Insecticides*, Marcel Dekker, 1971; and *Focus on Phytochemical Pesticides, Volume 1 (The Neem Tree)*, CRC Press, 1989. He also holds six U.S. patents on naturally occurring insecticides.

ROBERT L. METCALF, Professor Emeritus of Biology and Entomology and Research Professor of Environmental Studies, University of Illinois, Urbana-Champaign, is recognized internationally for his research on insect control, the chemistry and action of pesticides, and toxic substances in the environment. Among his achievements are the development of laboratory model ecosystem technology to screen pesticides for environmental acceptability and the discovery of carbamate insecticides and biodegradable substitutes for DDT. His work with the World Health Organization led to the development of insecticides for more effective control of vector-borne diseases. Professor Metcalf's recent research includes analyzing the effects of various industrial chemicals and pesticides on human health and environmental quality and investigating the coevolutionary and behavioral relationships between insect pests and cultivars, seeking new approaches to insect pest management. He was president of the Entomological Society of America in 1958 and has received numerous awards, including the Charles F. Spencer Award and the International Award in Pesticide Chemistry of the American Chemical Society, the Founders' Award of the Society for Environmental Toxicology and Chemistry, the Kenneth P. Dubois Award of the Society of Toxicology, the Memorial Lecture Award of the Entomological Society of America, and the Order of Cherubini from the University of Pisa. He is a member of the National Academy of Sciences and a fellow of the American Academy of Arts and Sciences.

RAMESH C. SAXENA, senior principal scientist, is the head of the Integrated Pest Management Section at the International Centre of Insect Physiology and Ecology (ICIPE), Nairobi, Kenya. He received his M.S. in tropical entomology from the University of Hawaii in 1966 and his Ph.D. in host plant resistance to insect pests from Delhi University in 1973. In 1975, he joined the International Rice Research Institute (IRRI) as a post-doctoral fellow in entomology. In 1977, he joined the ICIPE-IRRI project on major rice pests as an entomologist. From 1987 to July 1991, he served as entomologist in IRRI's Genetic Evaluation and Utilization Program. His major contributions include development of methodologies for efficient insect-rearing and screening of rice germplasm, including wild rices, biochemical plant-insect interactions, role of rice plant biotypes, and biointensive pest management. He conceptualized the relevance of botanical pest control for resource-limited farmers and demonstrated the potential of neem (*Azadirachta indica*) and other nonedible oil trees for ecologically sound pest management. He pioneered the introduction and large-scale planting of neem in the Philippines and Latin American countries. He also developed a simple process for extracting "neem seed bitters" for pest control. He has been an invited speaker at more than 40 international conferences and symposia and has published more than 200 scientific and professional articles. He was president of the Philippine Association of Entomologists in 1987–1988 and won several awards in the Philippines. His research work has been featured in international press releases and TV documentaries: "Coast-to-Coast" (Philippines), "Beyond 2000" (Australia), "State of the Earth" and "Discovery" (USA), and "Krishi Darshan" (India).

DAVID W. UNANDER, a plant breeder, has worked for the past five years at the Fox Chase Cancer Center in Philadelphia, Pennsylvania, on plants with activity against hepatitis B and other viruses. He also serves as an adjunct professor at Eastern College, St. Davids, Pennsylvania, where he teaches a course in appropriate technology through an M.B.A. program in international economic development. Previously he bred improved vegetables for the tropics as an assistant professor at the University of Puerto Rico. He received his B.S. and his M.S. in plant and soil science from Southern Illinois University in 1977 and 1979, and his Ph.D. in plant breeding from the University of Minnesota in 1983. He is a member and treasurer of the board of the Educational Concerns for Hunger Organization (ECHO), a nonprofit agency providing free extension information and experimental quantities of seeds on new crops and varieties to parties involved in international development. He has published extensively on the ethnobotany, cultivation, and biological activity of *Phyllanthus* species (Euphorbiaceae), as well as various articles relating to variety selection in pumpkins and squashes, vegetable peppers, soybeans, and dry beans.

NOEL D. VIETMEYER, staff officer and technical writer for this study, is a senior program officer of the Board on Science and Technology for International Development. A New Zealander with a Ph.D. in organic chemistry from the University of California, Berkeley, he now works on innovations in science and technology that are important for the future of developing countries.

The BOSTID Innovation Program

Since its inception in 1970, BOSTID has had a small project to evaluate innovations that could help the Third World. Formerly known as the Advisory Committee on Technology Innovation (ACTI), this program has been identifying unconventional developments in science and technology that might help solve specific developing country problems. In a sense, it acts as an "innovation scout"—providing information on options that should be tested or incorporated into activities in Africa, Asia, and Latin America.

So far, the BOSTID innovation program has published about 40 reports, covering, among other things, underexploited crops, trees, and animal resources, as well as energy production and use. Each book is produced by a committee of scientists and technologists (including both skeptics and proponents), with scores (often hundreds) of researchers contributing their knowledge and recommendations through correspondence and meetings.

These reports are aimed at providing reliable and balanced information, much of it not readily available elsewhere and some of it never before recorded. In its two decades of existence, this program has distributed approximately 350,000 copies of its reports. Among other things, it has introduced to the world grossly neglected plant species such as jojoba, guayule, leucaena, mangium, amaranth, and the winged bean.

BOSTID's innovation books, although often quite detailed, are designed to be easy to read and understand. They are produced in an attractive, eye-catching format, their text and language carefully crafted to reach a readership that is uninitiated in the given field. In addition, most are illustrated in a way that helps readers deduce their message from the pictures and captions, and most have brief, carefully selected bibliographies, as well as lists of research contacts that lead readers to further information.

By and large, these books aim to catalyze actions within the Third World, but they usually also have utility in the United States, Europe, Japan, and other industrialized nations.

So far, the BOSTID innovation project on underexploited Third-World resources (Noel Vietmeyer, Director and Scientific Editor) has produced the following reports.

Ferrocement: Applications in Developing Countries (1973). 89 pp.
Mosquito Control: Perspectives for Developing Countries (1973). 62 pp.
Some Prospects for Aquatic Weed Management in Guyana (1974). 33 pp.

Roofing in Developing Countries: Research for New Technologies (1974). 70 pp.
An International Centre for Manatee Research (1974). 34 pp.
More Water for Arid Lands (1974). 149 pp.
Products from Jojoba (1975). 30 pp.
Underexploited Tropical Plants (1975). 184 pp.
The Winged Bean (1975). 39 pp.
Natural Products for Sri Lanka's Future (1975). 53 pp.
Making Aquatic Weeds Useful (1976). 169 pp.
Guayule: An Alternative Source of Natural Rubber (1976). 77 pp.
Aquatic Weed Management: Some Prospects for the Sudan (1976). 57 pp.
Ferrocement: A Versatile Construction Material (1976). 106 pp.
More Water for Arid Lands (French edition, 1977). 148 pp.
Leucaena: Promising Forage and Tree Crop for the Tropics (1977). 110 pp.
Natural Products for Trinidad and the Caribbean (1979). 50 pp.
Tropical Legumes (1979). 326 pp.
Firewood Crops: Shrub and Tree Species for Energy Production (1980). 236 pp.
Water Buffalo: New Prospects for an Underutilized Animal (1981). 111 pp.
Sowing Forests from the Air (1981). 56 pp.
Producer Gas: Another Fuel for Motor Transport (1983). 95 pp.
Producer Gas Bibliography (1983). 50 pp.
The Winged Bean: A High-Protein Crop for the Humid Tropics (1981). 41 pp.
Mangium and Other Fast-Growing Acacias (1983). 56 pp.
Calliandra: A Versatile Tree for the Humid Tropics (1983). 45 pp.
Butterfly Farming in Papua New Guinea (1983). 33 pp.
Crocodiles as a Resource for the Tropics (1983). 52 pp.
Little-Known Asian Animals With Promising Economic Future (1983). 125 pp.
Casuarinas: Nitrogen-Fixing Trees for Adverse Sites (1983). 112 pp.
Amaranth: Modern Prospects for an Ancient Crop (1983). 74 pp.
Leucaena: Promising Forage and Tree Crop (Second edition, 1984). 93 pp.
Jojoba: A New Crop for Arid Lands (1985). 100 pp.
Quality-Protein Maize (1988). 100 pp.
Triticale: A Promising Addition to the World's Cereal Grains (1989). 103 pp.
Lost Crops of the Incas: Little-Known Plants of the Andes with Promise for Worldwide Cultivation (1989). 415 pp.

Microlivestock: Little-Known Small Animals with a Promising Economic Future (1991). 450 pp.
Neem: A Tree For Solving Global Problems (1992).
Vetiver Grass (1993). 141 pp.
Lost Crops of Africa: Volume 1—Grains (1994).
Lost Crops of Africa: Volume 2—Cultivated Fruits
Lost Crops of Africa: Volume 3—Wild Fruits
Lost Crops of Africa: Volume 4—Vegetables
Lost Crops of Africa: Volume 5—Legumes
Lost Crops of Africa: Volume 6—Roots and Tubers
Underexploited Tropical Fruits

Board on Science and Technology for International Development

Board on Science and Technology for International Development
Publications and Information Services (FO-2060Z)
Office of International Affairs
National Research Council
2101 Constitution Avenue, N.W.
Washington, D.C. 20418 USA

How to Order BOSTID Reports

BOSTID manages programs with developing countries on behalf of the U.S. National Research Council. Reports published by BOSTID are sponsored in most instances by the U.S. Agency for International Development. They are intended for distribution to readers in developing countries who are affiliated with governmental, educational, or research institutions, and who have professional interest in the subject areas treated by the reports.

BOSTID books are available from selected international distributors. For more efficient and expedient service, please place your order with your local distributor. Requestors from areas not yet represented by a distributor should send their orders directly to BOSTID at the above address.

Energy

33. **Alcohol Fuels: Options for Developing Countries.** 1983, 128 pp. Examines the potential for the production and utilization of alcohol fuels in developing countries. Includes information on various tropical crops and their conversion to alcohols through both traditional and novel processes. ISBN 0-309-04160-0.

36. **Producer Gas: Another Fuel for Motor Transport.** 1983, 112 pp. During World War II Europe and Asia used wood, charcoal, and coal to fuel over a million gasoline and diesel vehicles. However, the technology has since been virtually forgotten. This report reviews producer gas and its modern potential. ISBN 0-309-04161-9.

56. **The Diffusion of Biomass Energy Technologies in Developing Countries.** 1984, 120 pp. Examines economic, cultural, and political factors that affect the introduction of biomass-based energy technologies in developing countries. It includes information on the opportunities for these technologies as well as conclusions and recommendations for their application. ISBN 0-309-04253-4.

Technology Options

14. **More Water for Arid Lands: Promising Technologies and Research Opportunities.** 1974, 153 pp. Outlines little-known but promising technologies to supply and conserve water in arid areas. ISBN 0-309-04151-1.

21. **Making Aquatic Weeds Useful: Some Perspectives for Developing Countries.** 1976, 175 pp. Describes ways to exploit aquatic weeds for grazing and by harvesting and processing for use as compost, animal feed, pulp, paper, and fuel. Also describes utilization for sewage and industrial wastewater. ISBN 0-309-04153-X.

34. **Priorities in Biotechnology Research for International Development: Proceedings of a Workshop.** 1982, 261 pp. Report of a workshop organized to examine opportunities for biotechnology research in six areas: 1) vaccines, 2) animal production, 3) monoclonal antibodies, 4) energy, 5) biological nitrogen fixation, and 6) plant cell and tissue culture. ISBN 0-309-04256-9.

61. **Fisheries Technologies for Developing Countries.** 1987, 167 pp. Identifies newer technologies in boat building, fishing gear and methods, coastal mariculture, artificial reefs and fish aggregating devices, and processing and preservation of the catch. The emphasis is on practices suitable for artisanal fisheries. ISBN 0-309-04260-7.

73. **Applications of Biotechnology to Traditional Fermented Foods.** 1992, 207 pp. Microbial fermentations have been used to produce or preserve foods and beverages for thousands of years. New techniques in biotechnology allow better understanding of these tranformations so that safer, more nutritious products can be obtained. This report examines new developments in traditional fermented foods. ISBN 0-309-04685-8.

Plants

47. **Amaranth: Modern Prospects for an Ancient Crop.** 1983, 81 pp. Before the time of Cortez, grain amaranths were staple foods of the Aztec and Inca. Today this nutritious food has a bright future. The report discusses vegetable amaranths also. ISBN 0-309-04171-6.

53. **Jojoba: New Crop for Arid Lands.** 1985, 102 pp. In the last 10 years, the domestication of jojoba, a little-known North American desert shrub, has been all but completed. This report describes the plant and its promise to provide a unique vegetable oil and many likely industrial uses. ISBN 0-309-04251-8.

63. **Quality-Protein Maize.** 1988, 130 pp. Identifies the promise of a nutritious new form of the planet's third largest food crop. Includes information on the importance of maize, malnutrition and protein quality, experiences with quality-protein maize (QPM), QPM's potential uses in feed and food, nutritional qualities, genetics, research needs, and limitations. ISBN 0-309-04262-3.

64. **Triticale: A Promising Addition to the World's Cereal Grains.** 1988, 105 pp. Outlines the recent transformation of triticale, a hybrid between wheat and rye, into a food crop with much potential for many marginal lands. The report discusses triticale's history, nutritional quality, breeding, agronomy, food and feed uses, research needs, and limitations. ISBN 0-309-04263-1.

67. **Lost Crops of the Incas.** 1989. 415 pp. The Andes is one of the seven major centers of plant domestication but the world is largely unfamiliar with its native food crops. When the Conquistadores brought the potato to Europe, they ignored the other domesticated Andean crops—fruits, legumes, tubers, and grains that had been cultivated for centuries by the Incas. This book focuses on 30 of the "forgotten" Incan crops that show promise not only for the Andes but for warm-temperate, subtropical, and upland tropical regions in many parts of the world. ISBN 0-309-04264-X.

70. **Saline Agriculture: Salt-Tolerant Plants for Developing Countries.** 1989, 150 pp. The purpose of this report is to create greater awareness of salt-tolerant plants and the special needs they may fill in developing countries. Examples of the production of food, fodder, fuel, and other products are included. Salt-tolerant plants can use land and water unsuitable for conventional crops and can harness saline resources that are generally neglected or considered as impediments to, rather than opportunities for, development. ISBN 0-309-04266-6.

Innovations in Tropical Forestry

35. **Sowing Forests from the Air.** 1981, 64 pp. Describes experiences with establishing forests by sowing tree seed from aircraft. Suggests testing and development of the techniques for possible use where forest destruction now outpaces reforestation. ISBN 0-309-04257-7.

41. **Mangium and Other Fast-Growing Acacias for the Humid Tropics.** 1983, 63 pp. Highlights 10 acacia species that are native to the tropical rainforest of Australasia. That they could become valuable forestry resources elsewhere is suggested by the exceptional performance of *Acacia mangium* in Malaysia. ISBN 0-309-04165-1.

42. **Calliandra: A Versatile Small Tree for the Humid Tropics.** 1983, 56 pp. This Latin American shrub is being widely planted by villagers and government agencies in Indonesia to provide firewood, prevent erosion, provide honey, and feed livestock. ISBN 0-309-04166-X.

43. **Casuarinas: Nitrogen-Fixing Trees for Adverse Sites.** 1983, 118 pp. These robust, nitrogen-fixing, Australasian trees could become valuable resources for planting on harsh eroding land to provide fuel and other products. Eighteen species for tropical lowlands and highlands, temperate zones, and semiarid regions are highlighted. ISBN 0-309-04167-8.

52. **Leucaena: Promising Forage and Tree Crop for the Tropics.** 1984 (2nd edition), 100 pp. Describes a multipurpose tree crop of potential value for much of the humid lowland tropics. Leucaena is one of the fastest growing and most useful trees for the tropics. ISBN 0-309-04250-X.

71. **Neem: A Tree for Solving Global Problems.** 1992, 148 pp. The neem tree offers great potential for agricultural, industrial, and commercial exploitation, and is potentially one of the most valuable of all arid-zone trees. It shows promise for pest control, reforestation, and improving human health. Safe and effective pesticides can be produced from seeds at the village level with simple technology. Neem can grow in arid and nutrient-deficient soils and is a fast-growing source of fuelwood. ISBN 0-309-04686-6.

Managing Tropical Animal Resources

32. **The Water Buffalo: New Prospects for an Underutilized Animal.** 1981, 188 pp. The water buffalo is performing notably well in recent trials in such unexpected places as the United States, Australia, and Brazil. Report discusses the animal's promise, particularly emphasizing its potential for use outside Asia. ISBN 0-309-04159-7.

44. **Butterfly Farming in Papua New Guinea.** 1983, 36 pp. Indigenous butterflies are being reared in Papua New Guinea villages in a formal government program that both provides a cash income in remote rural areas and contributes to the conservation of wildlife and tropical forests. ISBN 0-309-04168-6.

45. **Crocodiles as a Resource for the Tropics.** 1983, 60 pp. In most parts of the tropics, crocodilian populations are being decimated, but programs in Papua New Guinea and a few other countries demonstrate that, with care, the animals can be raised for profit while protecting the wild populations. ISBN 0-309-04169-4.

46. **Little-Known Asian Animals with a Promising Economic Future.** 1983, 133 pp. Describes banteng, madura, mithan, yak, kouprey, babirusa, javan warty pig, and other obscure but possibly globally useful wild and domesticated animals that are indigenous to Asia. ISBN 0-309-04170-8.

68. **Microlivestock: Little-Known Small Animals with a Promising Economic Future.** 1990, 449 pp. Discusses the promise of small breeds and species of livestock for Third World villages. Identifies more than 40 species, including miniature breeds of cattle, sheep, goats, and pigs; eight types of poultry; rabbits; guinea pigs and other rodents; dwarf deer and antelope; iguanas; and bees. ISBN 0-309-04265-8.

Health

49. **Opportunities for the Control of Dracunculiasis.** 1983, 65 pp. Dracunculiasis is a parasitic disease that temporarily disables many people in remote, rural areas in Africa, India, and the Middle East. Contains the findings and recommendations of distinguished scientists who were brought together to discuss dracunculiasis as an international health problem. ISBN 0-309-04172-4.

55. **Manpower Needs and Career Opportunities in the Field Aspects of Vector Biology.** 1983, 53 pp. Recommends ways to develop and train the manpower necessary to ensure that experts will be available in the future to understand the complex ecological relationships of vectors with human hosts and pathogens that cause such diseases as malaria, dengue fever, filariasis, and schistosomiasis. ISBN 0-309-04252-6.

60. **U.S. Capacity to Address Tropical Infectious Diseases.** 1987, 225 pp. Addresses U.S. manpower and institutional capabilities in both the public and private sectors to address tropical infectious disease problems. ISBN 0-309-04259-3.

Resource Management

50. **Environmental Change in the West African Sahel.** 1984, 96 pp. Identifies measures to help restore critical ecological processes and thereby increase sustainable production in dryland farming, irrigated agriculture, forestry and fuelwood, and animal husbandry. Provides baseline information for the formulation of environmentally sound projects. ISBN 0-309-04173-2.

51. **Agroforestry in the West African Sahel.** 1984, 86 pp. Provides development planners with information regarding traditional agroforestry systems—their relevance to the modern Sahel, their design, social and institutional considerations, problems encountered in the practice of agroforestry, and criteria for the selection of appropriate plant species to be used. ISBN 0-309-04174-0.

72. **Conserving Biodiversity: A Research Agenda for Development Agencies.** 1992, 127 pp. Reviews the threat of loss of biodiversity and its context within the development process and suggests an agenda for development agencies. ISBN 0-309-04683-1.

74. **Vetiver Grass: A Thin Green Line Against Erosion.** 1993, 182 pp. Vetiver is a little-known grass that seems to offer a practical solution for controlling soil loss. Hedges of this deeply rooted species catch and hold back sediments. The stiff foliage acts as a filter that also slows runoff and keeps moisture on site, allowing crops to thrive when neighboring ones are desiccated. In numerous tropical locations, vetiver hedges have restrained erodible soils for decades and the grass—which is pantropical—has shown little evidence of weediness. ISBN 0-309-04269-0.

BOSTID Publication Distributors

U.S.:

AGRIBOOKSTORE
1611 N. Kent Street
Arlington, VA 22209

AGACCESS
PO Box 2008
Davis, CA 95617

Europe:

I.T. PUBLICATIONS
103-105 Southhampton Row
London WC1B 4HH
United Kingdom

S. Toeche-Mittler
TRIOPS Department
Hindenburgstr. 33
6100 Darmstadt
Germany

T.O.O.L. PUBLICATIONS
Sarphatistraat 650
1018 AV Amsterdam
Netherlands

Asia:

ASIAN INSTITUTE OF TECHNOLOGY
Library & Regional
Documentation Center
PO Box 2754
Bangkok 10501
Thailand

NATIONAL BOOKSTORE
Sales Manager
PO Box 1934
Manila
Philippines

UNIVERSITI OF MALAYA COOP. BOOKSHOP LTD.
Universiti of Malaya
Main Library Building
59200 Kuala Lumpur
Malaysia

RESEARCHCO PERIODICALS
1865 Street No. 139
Tri Nagar
Delhi 110 035
India

CHINA NATL. PUBLICATIONS
IMPORT & EXPORT CORP.
PO Box 88F
Beijing
China

South America:

ENLACE LTDA.
Carrera 6a. No. 51-21
Bogota, D.E.
Colombia

Africa:

TAECON
c/o Agricultural Engineering Dept
P.O. Box 170 U S T
Kumasi
Ghana

Australasia:

TREE CROPS CENTRE
P.O. Box 27
Subiaco, WA 6008
Australia

ORDER FORM

Please indicate on the labels below the names of colleagues, institutions, libraries, and others that might be interested in receiving a copy of Neem: A Tree for Solving Global Problems.

Please return this form to:

Neem Report, FO 2060V
National Academy of Sciences
2101 Constitution Avenue N.W.
Washington, D.C. 20418, USA

71

71

71

71

71

71